Many Visions, Many Aims
Volume 2

Many Visions, Many Aims

Volume 2

A Cross-National Investigation of
Curricular Intentions in School Science

WILLIAM H. SCHMIDT
Michigan State University, East Lansing, Michigan, U.S.A.

SENTA A. RAIZEN
The National Center for Improving Science Education

EDWARD D. BRITTON
The National Center for Improving Science Education

LEONARD J. BIANCHI
Michigan State University, East Lansing, Michigan, U.S.A.

and

RICHARD G. WOLFE
The Ontario Institute for Studies in Education, Toronto, Canada

KLUWER ACADEMIC PUBLISHERS
DORDRECHT / BOSTON / LONDON

A C.I.P. Catalogue record for this book is available from the Library of Congress.

ISBN 0-7923-4438-3 (Volume 2)
ISBN 0-7923-4442-1 (Set of 3 Volumes)
ISBN 0-7923-4439-1 (Volume 2 PB)
ISBN 0-7923-4443-X (Set of 3 Volumes PB)

Published by Kluwer Academic Publishers,
P.O. Box 17, 3300 AA Dordrecht, The Netherlands.

Sold and distributed in the U.S.A. and Canada
by Kluwer Academic Publishers,
101 Philip Drive, Norwell, MA 02061, U.S.A.

In all other countries, sold and distributed
by Kluwer Academic Publishers,
P.O. Box 322, 3300 AH Dordrecht, The Netherlands.

Printed on acid-free paper

Printed in the Netherlands

Table of Contents

Appendices (continued)

PREFACE

The Third International Mathematics and Science Study (TIMSS), sponsored by the International Association for the Evaluation of Educational Achievement (IEA) and the governments of the participating countries, is a comparative study of education in mathematics and the sciences conducted in approximately 50 educational systems on six continents. The goal of TIMSS is to measure student achievement in mathematics and science in participating countries and to assess some of the curricular and classroom factors that are related to student learning in these subjects. The study is intended to provide educators and policy makers with an unparalleled and multidimensional perspective on mathematics and science curricula; their implementation; the nature of student performance in mathematics and science; and the social, economic, and educational context in which these occur.

TIMSS focuses on student learning and achievement in mathematics and science at three different age levels, or populations.

- *Population 1* is defined as all students enrolled in the two adjacent grades that contain the largest proportion of 9-year-old students;

- *Population 2* is defined as all students enrolled in the two adjacent grades that contain the largest proportion of 13-year-old students; and

- *Population 3* is defined as all students in their final year of secondary education, including students in vocational education programs.

In addition, Population 3 has two "specialist" subpopulations: students taking advanced courses in mathematics (mathematics specialists), and students taking advanced courses in physics (physics specialists).

Population 2 was the core of TIMSS and sampling these students was required of all participating countries. Participating in the various aspects of TIMSS for Populations 1 and 3 (including the latter's subpopulations) was optional but each country was encouraged to participate in all options.

The design of the study and its instrumentation are based on a conceptual framework that focuses on the system, school, classroom, and individual student levels. TIMSS was guided by the following questions:

1. What are students expected to learn?

2. Who delivers the instruction?

3. How is instruction organized?

4. What have students learned?

Each aspect of the TIMSS project — the curriculum analysis, the context questionnaire surveys, and the student assessments — was designed to address one or more of these four focal questions.

What are students expected to learn?

The specification of goals at the national or regional level, which IEA has traditionally termed the "intended curriculum," defines what students are expected to learn. An analysis of science and mathematics textbooks and curriculum guides used in the participating countries, along with a questionnaire administered to subject matter experts, provided data describing what students in the participating countries are expected to learn according to national or regional educational goals. Results from a preliminary analysis of early curriculum data were used to create the TIMSS curriculum frameworks and to select the topics to be included in the achievement tests. The curriculum analysis, a portion of which is presented in this volume, will also provide invaluable information for interpreting the achievement results.

Who delivers the instruction? How is instruction organized?

Questionnaires were designed to address these questions as well as to provide information on the hypothesized correlates of achievement and about the social and economic contexts in which learning is taking place.

The participation questionnaire provided contextual information about a country's educational system. The first questionnaire dealt with the structure of the school system, student ages and grades, and the school calendar. The second questionnaire focused on school systems and curricular/instructional decisions, teaching qualifications, and course structures.

In the TIMSS main survey, context questionnaires are administered to three separate groups:

1. The students sampled for TIMSS assessment;

2. The mathematics and science teachers of the selected students; and

3. The principals of the schools containing the selected students.

A student background questionnaire was administered to every student who participated in the achievement assessment. This provided information on student attitudes toward mathematics and science, personal and academic background, school experience, and career/educational goals. A teacher questionnaire was administered to the teachers of Population 1 and 2 students only. It provided information on teachers' academic and professional backgrounds, their instructional practices, pedagogical beliefs, and attitudes toward mathematics and science as well as their evaluation of their students' curricular opportunities and their own curricular goals for students. Finally, a school questionnaire was completed by the appropriate school administrator in each participating school. It provided data on school staffing and resources, textbook policy, curricular policy, and characteristics of the student body and the school's community.

What have students learned?

Assessment of student achievement in key mathematics and science topics at each of the focal student populations entails both multiple-choice and open-response items (short answer and extended response). The tests were constructed from specifications reflecting TIMSS curriculum frameworks and include items developed and reviewed by international experts. At Populations 1 and 2, a subpopulation of sampled students participated in the performance assessment component of TIMSS. This consisted of a range of hands-on mathematics and science tasks. The tasks were set up in stations by assessment staff. Each station consisted of one or two tasks in mathematics, science, or a combination of both, with students attending three different stations with 30 minutes at each station to complete the designated activities.

The present volume is the second report to come from the international curriculum analysis. The first volume considered the mathematics curriculum for students across the many countries participating in TIMSS. The current volume considers the sciences curriculum intended for students. More detailed volumes that look specifically at textbooks for mathematics and the sciences will also be produced. These curriculum volumes should prove to be an invaluable resource for understanding and interpreting the results of the student assessments.

As is common with any research, there is a lag between the time data are collected and results are published. Most of the data presented represent what was available in countries in 1990 and 1991 (with some data collected in 1992 and 1993). Some things have, no doubt, changed since then, and throughout the book the data are discussed with this in mind.

Finally, the work reported here would not have been possible without the support of grants RED 9252935, SED 9054619, and RED 9550107 from the National Science Foundation (NSF) in conjunction with the National Center for Education Statistics (NCES). NSF and NCES provided the funds for this resarch, but do not assume responsibility for the findings or their interpretations.

I. Characterizing Curriculum: An Overview

Chapter 1

SCIENCE CURRICULUM: INVESTIGATING CURRICULAR INTENTIONS

A 9-year-old girl sits in her classroom in South Africa looking through the window — earlier this week her teacher conducted the class in building a bird feeder. The girl now observes birds visiting the feeder and records her observations on a table that she has copied from her environment studies textbook. Earlier, a group of 13-year-old pupils in the Philippines followed the directions in their science and technology textbook for the construction of a model steam turbine. After completing this activity, they answered some questions on its operation in their textbook. In Spain, other 13-year-old students complete exercises in their textbook solving different chemical reaction equations for the time it takes for each reagent to disappear given the time it takes the new compound to form. After school, their teacher joins with several colleagues to prepare further materials for next week's classes, sharing their ideas with each other and reading a set of detailed suggestions provided by the Ministry of Education to guide their teaching. In Germany, that evening, a teacher reviews her notes for the physics class she will teach the next day for students in their final year of secondary school and compares them to the latest version of the official document guiding what she should present. Meanwhile, one of her students reviewed the notes he took during today's class describing the photoelectric effect, a topic he must master to do well on his Abitur[1] examination later this year. He compares his notes to the examples and problems in the textbook he purchased the previous August.

What do all these situations and others like them have in common? All involve children. All involve the sciences. All involve curriculum materials that students or teachers use in some way during science instruction. Less obviously, all involve plans on the part of the teacher and, at least indirectly, curriculum guides that help determine teachers' practices. Many factors shape these strategies — experiences and personal beliefs about what is important in the sciences, views of how the sciences should be approached in the classroom, official guidelines stating curricular goals, student textbooks, or some combination of these and other factors. These learning situations all flow from the visions and intentions inherent in that system's science educational policies.

How are all these situations different? Most obviously, they differ in the sciences studied. Even when dealing with the same topic, they use different activities and materials to approach and explore the topic with children. They differ in what was covered previously and in what students are expected to master at that time. They vary in how long lessons on that topic will continue, and in many other ways. The situations differ in *how much* class time is spent on a given subject, and *how* that time is spent. How free teachers are to put their personal stamp on individual lessons also varies. The use of science textbooks, if any, varies. Preparation differs in how explicitly official documents direct teachers in providing instruction. The situations differ in the experiences science curriculum planners intended that children share. They also vary in how they will be judged — in how the system will determine whether teachers and students successfully attained intended goals.

Across the world, teachers and their students engage in the study of the sciences as they have for generations. It is widely believed that science is objective and universal, transcending national and cultural differences. But what students and teachers experience is not actual *scientific investigation* but *school science*. School science is science as it is conceptualized, represented, structured, and sequenced to share with the next generation through the common experiences of schooling. Like all of education — which is a common means of transmitting a nation's culture to its next generation — school science is profoundly cultural. This monograph aims to explore this cultural aspect of science education by attempting, through an investigation of curricula, to identify differences and similarities that underlay the curricula of school science of many countries.

Why Investigate Science Curricula?

Each day in classrooms around the world, children and their teachers engage in some aspect of learning science. These daily educational experiences, which seem so very ordinary, are actually guiding and shaping children's lives. *The cumulative effect of the experiences pupils share help mold them and help determine the course and quality of their lives.* Any set of science educational experiences will have some things in common (and many things that vary) among countries. Some differences are deep; some are incidental. Separating the two is essential to understanding better how children are educated in science. It is essential for seeking — within given cultural settings — more effective educational practices in schools, classrooms, and textbooks. *Curriculum is the most fundamental structure for these experiences.* It is a kind of underlying skeleton that gives characteristic shape and direction to science instruction in educational systems around the world.

These experiences do not occur randomly or accidentally. Instead, they are deliberately shaped based on visions of what education should be, ideas of how to create the formative experiences of education, and intended patterns of opportunities that organize the potential for those experiences. The myriad details of classroom life often hide the fact that instructional activities are intended to be planned and orderly implementations of the aims and intentions of educational authorities and curriculum designers. The plan that expresses these aims and intentions, that takes them from vision to implementation, that serves as the broad course that runs throughout formal schooling, is *curriculum*. Curriculum provides a basic outline of planned and sequenced educational opportunities. It is "the idea of structure imposed by authority for the purpose of bringing order to the conduct of schooling."[2]

The status and future of any nation are caught up in its children's lives and their education. No nation can thrive for long if it seriously fails to educate its children. Thus, no nation can afford to ignore curriculum — for the sciences or any other important subjects. Curriculum must be understood, valued, and deliberately guided. In forming educational policy, educational systems help shape the future of their children and their country.

In recognition of the central importance of curriculum in education, the Third International Mathematics and Science Study (TIMSS) includes large-scale cross-national analyses of science curricula as part of its examination of science education and its attainment in almost 50 nations. The TIMSS curriculum analysis investigates curriculum documents and the visions and aims on which they are based.

Earlier Studies

Researchers have long been interested in curricular intentions, yet few have examined them and their relationships to national strategies for covering science topics. The little work that has been done points to the potential of future studies in this area for helping understand why science students' achievements and teachers' instructional methods differ among countries.

- For example, one study of national curriculum guides from a large multi-country sample revealed important regional differences in emphasis given to science instruction in primary and secondary school curricula.[3] This study found that countries in Eastern Europe and Latin America devoted significantly more time to the study of the sciences than did countries in the rest of the world.

- Some studies of European curricula have revealed considerable variation between countries in the degree to which physics is taught with a consideration of the interactions of applied physics and society.[4]

- Significant cross-national variation has also been noted in whether the scientific disciplines are taught as collections of concepts and facts or whether there is more emphasis on practical activities.[5]

- Even between textbooks series in the same country, there appear to be many significant differences when examined in detail. Such differences have been reported, for example, in the development and presentation of important concepts that lead to an understanding of biological evolution.[6]

The intriguing differences across countries suggested by these studies have clear implications for understanding how science curricula shape classroom activities and determine what skills students are likely to acquire. These studies illustrate some of the potential in investigating curricular documents.

The present curriculum analysis is a natural extension of the informal analyses of curriculum guides and textbooks found in earlier studies of the International Association for the Evaluation of Educational Achievement (IEA). In IEA's Second International Science Study, for example, researchers were asked to rate various science content areas according to whether or not (and the extent to which) they were covered in the curriculum of their country. Many participants assigned these ratings by consulting curriculum guides and textbooks.[7] This TIMSS study builds upon that idea but has developed more thorough, rigorous, data-gathering methods to document what countries intend to teach.

Curriculum and Student Experience

For about a century, curriculum has been understood in terms of student experiences in school. For example, Bobbitt (1918)[8] considered curriculum to be "the entire range of experiences, both directed and undirected, concerned in unfolding the abilities of the individual." He also called curriculum the *"series of things which children ... must do and experience* by way of developing abilities to do the things well that make up the affairs of adult life" (emphasis in original). Before Bobbit, John Dewey [9] had pointed out that the curriculum is the reflective formulation of experience, helping to guide an orderly succession of experiences for pupils in school. In this sense, science curricula are those experiences needed to develop students' science abilities.

Students' classroom experiences are created by the interaction of assigned tasks, activities to accomplish those tasks, materials supporting the activities, students, and teachers.[10] Each such interaction creates an experience for each student. Each experience is personal and idiosyncratic for each student. It is these individual experiences that are the basics of schooling and curricular experience. The variety and accumulation of these experiences for a student produce his or her educational attainments. Curriculum and teaching consistently reflect the desire to create sequences of common, shared experiences that result in desired attainments for as many students as possible.

Differences in children's curricular experiences produce individual and characteristic national outcomes in science learning. In this sense, experiences reflect educational intention and planning. These experiences, however, are also *personal*. Besides being the result of interactions with external factors — such as tasks, materials, and activities — they also involve internal factors — including each student's perceptions, interests, and motivations surrounding the phenomena experienced, involvement with the setting, and the culture and traditions that help shape those perceptions. While these sets of experiences and their cumulative effect are the objects of curricular intentions and planning, they can, since they are personal and individual, only be influenced partially and indirectly. *Thus, curricular intentions can at best only determine potential student experiences — experiences intended to be common for large numbers of students and to be causally linked to desired attainments.*

Curricula create and distribute educational opportunities in an effort to create desired common experiences. The characteristics of any particular science curriculum are thus captured in part by documenting aspects of the opportunities it creates. For further discussion of this view of curriculum, see appendix A; for information on a general model of potential experiences, see appendix B.

An Approach to Curriculum Analysis

The TIMSS curriculum analysis effort set out to understand the intentions, that is, the visions and aims, helping shape science curricula in some 50 nations. But how can curricular visions and their associated curricula be analyzed? In smaller scale projects, the "artifacts" of curricular visions are commonly used — curriculum guides,[11] textbooks, tests, and materials. Although these data provide only a partial glimpse of the curricular vision behind the everyday realities of educational life, they do allow certain inferences to be made about curricular intentions.

Before TIMSS, curricular documents had never been used as the primary data source for a project of this scale. Resource limitations caused previous large-scale curriculum studies to rely almost solely on expert opinions rather than curricular documents. While TIMSS also suffered from resource limitations — for example, it was not possible to translate all documents into a single common language — methodologies, described in the next section, were developed to enable the study to investigate curricular intention through careful document analysis supplemented by expert opinion.

Ideally, a full study of curricular documents would include ministerial policy documents, curriculum guidelines, course syllabi, textbooks, syllabi for national examinations, teacher pedagogical plans as they interpret broader requirements, tests, and so on. We limited our attention, however, to those central documents common to virtually all the participating countries - science curriculum guides and student textbooks.

- The *curriculum guides* are official documents that most clearly reflect the intentions, visions, and aims of curriculum makers.

- *Textbooks* provide less official, partial reflections of intentions as interpreted by other participants in the curriculum process. In many countries, textbooks capture important insights about intentions and aims.

Some expert opinion data supplement analyses of these documents. Supplemental materials and annotated teacher editions of textbooks would have offered additional data, but, because they were less consistently used, we did not include them in this study.

Analytic Methods

The TIMSS Science Framework

The first methodological requirement was to develop a tool for segmenting and categorizing pieces of the curricular documents and TIMSS tests. To this end, a unified set of categories was developed: the TIMSS science framework. The framework focuses on three aspects of science education and testing:

- *Content* refers simply to the science topic area.

- *Performance expectation* refers to common aspects of students' completing science tasks. This aspect specifies what the student is to do with the science content. Expectations were considered less culturally determined than postulating the cognitive processes of performing specific science tasks.

- *Perspective* refers to broad themes that might underlie many tasks, for example, emphasizing the societal importance of the sciences through specific contents or tasks.

Appendix C discusses the science framework in some detail and lists the categories of the framework's three aspects. Note that throughout this monograph, we use single quotation marks to indicate verbatim framework category labels: thus, 'biochemistry of genetics' is a category label. Informal topic areas and references to topics that are part but not all of a formal category are not enclosed in quotation marks.

Topic Trace Mapping

Some curricular information is essential for all grade levels. It was decided to obtain these data for all content topics in the science framework. Unfortunately, this was feasible only by using expert opinions rather than directly by document-based methods (except for a few science topics covered in depth).

Each country was asked to designate a panel of experts that would provide information on each grade in which each topic was covered. The experts were asked to base their responses on curriculum guides and other curricular materials, but the ways in which this task was done were not directly monitored.

For each topic, the experts were asked to provide two kinds of information. First, they were to indicate all grades in which aspects of that topic were covered. Second, they were asked to indicate in which grade the topic was the "focus" of more extensive or intensive curricular attention. Experts provided similar data for all the different major regions or streams in their country with differing science curricula. These coverage and focus data were collected for science content topics only; pilot work showed that this method did not provide reliable data on performance expectations or perspectives.

The result of this collection effort was a "trace" or "map" of the way in which each content topic entered and left the science curriculum across years of schooling. Any special focus or curricular attention was also mapped. To ensure reliability, countries were asked to provide their data more than once, asked to review the data they had provided, and asked direct questions about any apparent discrepancies. A more document-based method was used to gather fuller, in-depth information on a small, central set of topics; this technique was called "in-depth topic trace mapping." Those data are not reported in this monograph. See appendix D for a more detailed discussion of topic trace mapping.

Document Analysis

To specify science curricular intentions completely for any country would have required collecting information from every document — every curriculum guide and every textbook — for every science course at each grade and for each educational track or stream in the country. Such a vast amount of data collection was neither practical nor economically feasible in a single country, let alone in the almost 50 TIMSS countries. Instead, we decided to sample the documents.

To develop a practical sampling scheme and methodology, researchers had to balance several important, conflicting demands:

- The need for some information on intentions at all grade levels;

- The need for detailed information at some grade levels, particularly at the focal grades at which achievement testing would occur; and

- The need to be sensitive to (1) important national differences in educational governance and textbook role and (2) curricula and textbook differences within countries.

The grades selected for document collection were those for which TIMSS achievement testing would take place: in each country, the two consecutive grades with the majority of 9-year-olds (hereafter called TIMSS Population 1), the two grades with the majority of 13-year-olds (Population 2), and the physics specialist courses in the final year of secondary school (Population 3). Documents were sampled rather than surveyed exhaustively.

Each national document sample required selecting curriculum guides and textbooks to represent those used for at least half of the students in the targeted grades. Major regional, school type, and other strata were to be represented appropriately in the sample. National curriculum guides were sampled if they existed; otherwise, the sample consisted of appropriate regional, provincial, etc., guides for each grade. Similarly, official national textbooks were sampled if they existed; otherwise, the most widely used commercial textbooks were selected. The documents included for each country were chosen by country representatives in consultation with TIMSS. Documents were marked for analysis in each country and coded (without translations) for the three aspects of the TIMSS science framework in the individual countries. Next, informational questionnaires in English were provided, and marked archival copies of the documents were sent to the curriculum analysis center for filing and reference.

Forty-eight countries reported at least some document-based curriculum data. These included 77 science curriculum guides for the upper grade of Population 1, 111 for the upper grade of Population 2, and 63 for the physics specialists of Population 3. The textbook sample comprised 96 science textbooks for the upper grade of Population 1, 156 for the upper grade of Population 2, and 60 for the physics specialists of Population 3. Thus, the analysis is based on a sample of 251 science curriculum guides and 312 textbooks, a total of 563 curriculum documents.

The documents were written in more than 30 different languages. Even for the sample, it was not feasible to translate complete documents. Instead, procedures were developed and shared that allowed each country to use its own team to mark and code important features of each document. The features chosen are discussed below. They certainly did not exhaust all aspects of the documents, but they did represent features that pilot work and early field trials suggested were important for capturing document intent, structure, and some essential content.

Many documents were quite long, so dividing them into smaller segments was essential. This was done in two stages. First, each document was separated into a series of *units* to capture major structures. Units were of several types, and each unit had its type determined and recorded by the appropriate country team. To capture even more finely grained structures in the documents and to attach framework codes to small, narrowly defined segments, each unit was further subdivided into one or more segments called *blocks*. For a listing and detailed description of the unit and block types coded, see appendix E.

Next, category codes for science content, performance expectations, and perspectives were assigned to *blocks*. These data — a "statistical abstract" of each document — were recorded on appropriate forms and cross-referenced to marked and numbered segments of the original documents. The marked documents, coding forms, and supplemental questionnaires for each document, unit, and block were sent to the curriculum analysis center for data entry and archiving.

These methods involved considerable decision making by each country's document analysis team. Several measures were taken to ensure uniformity and reliability. Detailed manuals were prepared for every procedure. Face-to-face training was provided to representatives of every country through a series of regional training sessions lasting almost 3 days each; for this training, a manual, examples, and exercises were prepared. An initial quality assurance phase required national teams to code a selected random sample of units, which were then translated and coded by international referees; failures to meet satisfactory criteria were noted and analyzed. Only when the criteria had been met (by further practice, training, coding new samples, etc.) were countries authorized to begin the main analysis. All received materials were carefully examined, and discrepancies between documents and forms were pursued individually to obtain clarifications. Finally, after the collection was complete, a further sample of submitted materials was selected, translated, and coded for each country. An appropriate reliability estimating procedure was developed and applied. In virtually every case, concordance of more than 80 percent (and often more than 90 percent) resulted. For more detail and examples of document analysis, see appendix E.

NOTES

[1]   In Germany, four examinations are given to students who wish to enter the university as part of the requirement for the *Abitur* certificate.

[2]   Jackson, P. W. 1992. "Conceptions of Curriculum and Curriculum Specialists." In P.W. Jackson, ed., *Handbook of Research on Curriculum*. New York: Macmillan Publishing Co.

[3]   Kamens, D. H., and Benavot, A. 1991. "Elite Knowledge for the Masses: The Origins and Spread of Mathematics and Science Education in National Curricula," *American Journal of Education* 99 (February): 137-180.

[4]   Millar, R.H. 1981. "Science Curriculum and Social Control," *Comparative Education* 17:23-30.

[5]   Holbrook, J.B. 1991. *The Relationship Between Curricula and Assessment.* Mimeograph. Paris: UNESCO.

[6]   Jeffrey, K.R., and Roach, L.E. 1994. "A Study of the Presence of Evolutionary Protoconcepts in Pre-High School Textbooks," *Journal of Research in Science Teaching* 31 (May): 507-518.

[7]   Rosier, M. J. 1987. "The Second International Science Study," *Comparative Education Review* 31 (February): 106-128.

[8]   Bobbitt, J. F. [1918] 1972. *The Curriculum.* New York: Arno Press, 42-43.

[9]   Dewey, J. 1902. *The Child and the Curriculum.* Chicago: University of Chicago Press.

[10]   Note that we focus here only on students' experiences occurring through and created by the deliberate efforts of schooling, since our goal is to investigate curriculum as intentional and planned.

[11]   "Curriculum guide" is used here as a generic, standardized term for countries' official documents setting out policies and curricular intentions. Curriculum guides provide an official statement of what a science curriculum is intended to be in a specific context.

Chapter 2

OVERVIEW OF RESULTS

This monograph presents the major findings of a large-scale investigation of the science curricular visions and aims of the almost 50 countries participating in the Third International Mathematics and Science Study (TIMSS). For the first time ever in a study of this scope, national curricular documents — specifically, official curriculum guides and textbooks — were the primary data source; they were supplemented with expert opinion. The analyses described in chapters 3 through 9 of this monograph are the first attempts to compare and contrast similarities and differences in science curricular intentions across the TIMSS countries. Much more work remains to be done with these data, and we suggest future research directions in chapter 10.

Monograph Organization

Given the scope of this investigation, the findings are organized for use by the educational policy makers, science education policy makers, science curriculum developers, teachers, and other professionals interested in effective science curricula who are the readers of this monograph. The monograph is organized into five major sections.

- **Characterizing Curricula: An Overview.** This section highlights the study's most important findings and caveats. Readers interested in a general introduction to and overview of the TIMSS curriculum analysis effort should read chapters 1 and 2.

- **Curricular Organization and Control.** The two chapters in this section — 3 and 4 — address general topics in comparing educational systems, their control, and related policy issues.

- **Reflections of Curriculum.** This section comprises chapters 5 through 9; it covers specific issues of science curricula and related policy. In general, chapters 5 through 8 look at broad similarities and differences in science content, performance expectations, and perspectives across the TIMSS countries. Chapter 9 presents a different focus on the data: here, the search is across smaller groupings of countries seeking common patterns in curricular intentions.

- **Consequences of Curriculum: Policy Implications.** This section looks at the implications of the curriculum analysis findings. Chapter 10 summarizes broad findings and points out directions for future studies.

- **Supporting Data.** The several appendices to this monograph present more detailed versions of data and information appearing in the main text. Specifically, the first five appendices (appendices A through E) focus on key models and methods underlying the investigation. Appendices F through H list the various resources on which this work relied. Appendix I contains comprehensive data tables and figures.

Monograph Conventions

We have relied on several conventions in presenting the information in this monograph, specifically:

- **Sampling in Graphic Displays.** Throughout the monograph, tables and figures contain representative, illustrative results rather than complete data (which, for most tables, are included in appendix I). In some displays, specific countries, topics, or grade levels were selected, either according to particular criteria or to provide a typical picture. When particular criteria are involved, it is stated so in the discussion. Otherwise, countries and topics were selected to provide a representative illustration of the patterns and variations in the data. Some effort was made to rotate the selection of representative countries so that each TIMSS country appears in at least some of the figures and tables. Unless specifically noted, nothing else should be read into the selection of particular countries or topics.

- **TIMSS Focus Populations.** TIMSS is designed to gather information about mathematics and science learning and achievement for three different student populations. Student Population 1 is defined as all those enrolled in the two adjacent grades that contain the largest proportion of 9-year-old students. Population 2 is defined as all students enrolled in the two adjacent grades that contain the largest proportion of 13-year-old students. Population 3 is defined as all students in their final year of secondary education. This definition includes two subgroups: "generalists," which includes all students including those in vocational programs, and "specialists," which includes those students taking advanced courses in physics (physics specialists). The focus in the curriculum analysis and the student assessment aspects of TIMSS has been upon those students in the upper grade of student Populations 1 and 2. For example, in a country in which most 9 year-olds were in grades 3 and 4, the TIMSS focus would be on students in grade 4.

- **TIMSS Framework Topics.** We have relied on the science framework developed by many persons, presented in several reports, and finally published in Robitaille et al. (1993) and summarized here in appendix C, in coding and describing curricular content, performance expectations, and perspectives. Where a verbatim curriculum framework topic label is used in this monograph, we have indicated it with single quotation marks (e.g., 'biochemistry of genetics'); informal topic areas and references to topics that are part but not all of a formal category are not enclosed in quotation marks.

Study Findings

Similarities Within a Context of Differences (See Chapter 3)

National and subnational education systems differed in several important ways. They differed in broad system features such as school entry age, the number of years of schooling provided, and the number of years of schooling that were compulsory. They also differed in how school grades were clustered — that is, in the grade levels at which breaks between elementary school, lower secondary school, and secondary school occurred. These differences affected the context of schooling, the resources available, and other aspects of the learning environment.

Educational systems also differed in their curricular organization. Even within broadly similar systems, there was considerable differentiation into various streams within schooling and various science curricula intended to serve different student cohorts. This clearly implies differences in the kinds of learning opportunities provided, in the science contents involved, in the typical expectations for students, and in the organizing and sequencing of the opportunities provided. The systems also differed with regard to how curricular decision making was distributed within the educational system — among ministries, schools, teachers, and so on. Which decisions were made by whom varied both among countries and among various national subsystems within many countries. Nations also differed in the complexity of their national systems. Some were relatively homogeneous, with essentially one centralized national system; others were more heterogeneous, with many subsystems and curricular and organizational differentiation even within these subsystems. For example, the United States is a "federalism" of more than 50 educational systems.

Any similarities among science curricula and education had to be sought within this context of differences among educational systems. Broad similarities in science education were not hard to find. All countries required science study for at least some grades. Of the topics in the TIMSS science framework, about 30 to 40 were introduced in every country at some point within schooling. All countries reported some interest in science education reform of one sort or another.

When analyses were made more specific, the patterns of national and subnational differences in science education persisted. The numbers of courses offered at various points in any stream of schooling — and the organization of those courses — varied. Curriculum guides, the official documents that set out science curricula, varied in their status, structure, size, specificity, and use. Textbooks also varied in their status, structure, size, and use.

Curriculum guides and textbooks from the participating countries also differed in the emphasis they placed on and the space they gave to various broad science topics. They differed in terms of what students were expected to be able to do with those topics. The sequences of content presentation varied among the textbooks; this was true both among countries and within some countries in which more than one text per grade level was used.

Differences among countries were hardly surprising, although their extent and the detailed level at which they persisted might surprise some researchers. Also, some broad similarities across countries were anticipated. However, what stands out in this initial survey is that the similarities in science curricula were found in a context of wide, detailed, and persistent differences and variations both among and within countries.

This is not to suggest that an empirical search for substantive similarities in intentions for science curricula would be fruitless. Rather, it emphasizes the care needed in such a search and reinforces the idea that, overall, any cross-national comparisons of science curricula require careful interpretation and contextualization when drawing inferences.

The Role of Science Curriculum Guides (See Chapter 4)

Science curriculum guides differed in their status, authority, and function among TIMSS countries. While always setting out official goals and aims for science curricula, curriculum guides in some countries clearly had the force of *binding* documents strongly guiding curricular implementation in instruction. In other countries, however, they seemed to have been taken more as suggestions — articulating goals and emphases, and directing choices in which textbook materials were used among those available. In the latter cases, individual teachers, groups of teachers, schools, and groups of schools had considerable decision-making authority regarding content emphases and instructional materials. Further, textbooks in those countries served at times almost as de facto science curricula, with curriculum guides helping to shape omissions, time allocations, and instructional goals.

Curriculum guides differed in their names ("frameworks," "syllabi," and "courses of study" were some common alternatives). They differed in their level of inclusiveness — that is, in the number of grades and ages to which any one document was relevant. These differences occurred both among countries and within some countries.

Curriculum guides also differed in their balance between more prescriptive approaches (for example, use of policy statements, objectives, and content specifications) and more facilitative approaches (such as pedagogical suggestions, examples, and assessment suggestions). Almost all curriculum guides used a combination of these approaches. They differed, however, in the relative use of prescriptive versus facilitative approaches; few countries evenly divided their use of these approaches.

Science content was specified in almost every part of most of these documents and for every country. Overwhelmingly, the content was indicated as presented at the most specific level made possible by the TIMSS science framework. Student performance expectations were somewhat less pervasive and less specific. These expectations were indicated by national coders as present in most sections of most curriculum guides but not in all. Often, these expectations were not marked at the most specific level available; this suggests more general statements of expectations in the documents, a mismatch between the possibilities in the framework and those needed to characterize the document, or both.

Although similarities were again in a context of differences, science curriculum guides overall seemed to function consistently to define and communicate curricular intentions and goals, to do so in a variety of ways, and to be specific about the science contents involved (persistently) and about what was expected of students (often). Certainly, these documents served in at least some countries as an important means of controlling and guiding curricular implementations in science instruction.

The Flow of Science Curricula (See Chapter 5)

Science curricula are implemented across the years of schooling. In that sense, any one curriculum has a "flow" across grades: Topics are introduced, continue for a time, and cease to receive attention. There were considerable national variations in flow. Primarily, the same topic was introduced at different times in different countries, and the coverage of a topic continued for different numbers of grades in different countries. Even so, it was possible to identify sets of topics that were characteristic — but not universally applicable — of groups of adjacent grades in many countries:

- The science curriculum in grades 1 through 3 was primarily devoted to taxonomy in the life and earth sciences. 'Weather and climate' was also common.

- In grades 4 through 6, life and earth science topics were extended and additional physics topics were commonly introduced. 'Pollution' and resource conservation concepts were introduced for the first time.

- In grades 7 and 8, many more physics and chemistry topics were introduced.

- Grades 9 and beyond, commonly saw the introduction of more advanced topics in genetics, physics, and chemistry.

Similar commonalities and differences occurred with regard to how long topics continued to be covered and when their coverage was completed. Countries differed in the number of topics intended for instruction in each grade, in how often and when new topics were introduced, and in how often and when old topics were dropped from continued curricular attention. Consequently, some countries had more focused curricula with attention restricted to a relatively small number of topics; others had comparatively more diverse curricula. Some countries had school science that was continually changing — new topics added, old topics dropped — while other countries made comparatively fewer changes.

The data on the flow of science curricula across the grades painted a rich picture of national differences but with surprising commonalities in the sequences of content coverage and in the places in which topics were in the curricular flow of many countries. These data represent important contextual information for interpreting science attainments.

Milestones in Science Curricula (See Chapter 6)

Science curricula do not flow across the grades as even, undifferentiated sequences. Each curriculum has certain milestones — key topics that are the focus of special curricular attention — scattered throughout their sequence. Countries supplied data indicating at which grades certain topics were the focus of special curricular attention. As always, there were many differences among countries and among science topics. Also, while data on commonly covered topics across grades show that many topics received curricular attention from many countries, data on focus topics show that far fewer topics received such attention from many countries. In fact, no topic at any grade level was indicated as a focus in the same grade for more than 70 percent of the countries. Such lack of common focus and marked cross-national diversity in curricular attention make science attainment testing difficult and require that the results of such testing be carefully interpreted.

In identifying milestone contents at key grades — the upper of the two grades for Population 1 (those containing the most 9-year-olds), the upper of the two for Population 2 (those containing the most 13-year-olds), and the physics specialist subpopulation of Population 3 (final year of secondary school) — data from the analysis of curriculum guides and textbooks can be used to identify common milestones. A portrait of commonly intended topics at these key grade levels (those present in the curriculum guides of 70 percent of the countries, the textbooks of 70 percent of the countries, or both) emerged:

- The upper grade of Population 1 was generally focused on taxonomy in earth and life science. In addition, some elements of physical and earth sciences were present.

- The upper grade of Population 2 showed great diversity, with topics from many areas in the science framework present.

- The physics specialists of Population 3 were usually exposed to a curriculum devoted to 19th century physics.

With these data, it was possible to compare what was commonly intended in curriculum guides with what was commonly intended in textbooks. Often, the same topics were widely present in both. Some topics — usually more advanced topics or those related to recent reforms in school science — were present in curriculum guides but not in textbooks. This suggested that more conservative approaches were taken in textbooks, or that there was a lag between when topics began to be emphasized and when they found their way into textbooks. Conversely, some topics — usually those already considered in previous grades — persisted in textbooks even when not widely present in curriculum guides. This further suggests more conservative visions of science curricula implicit in textbooks.

Topics widely present in both curriculum guides and textbooks (in 70 percent of the countries) and also emphasized in textbooks (indicated for at least 6 percent of the blocks in textbooks covering the topic, a cutoff empirically determined to indicate significantly more textbook emphasis for a topic) were identified. Very few topics were widely present and emphasized in this way.

- In the upper grade of Population 1, only two topics — 'plant, fungi types' and 'animal types' — were thus emphasized, but these two together accounted for only about a quarter of that population's textbook blocks.

- Only one topic was emphasized in this way at the upper grade of Population 2 — 'organs, tissues.' Far fewer of the textbook blocks (under 10 percent) were accounted for by this topic.

- For the Population 3 physics specialists, six topics were emphasized: 'energy types, sources, conversions,' 'wave phenomena,' 'light,' 'electricity', 'magnetism,' and 'time, space, motion.' 'Electricity' accounted for about 20 percent of these specialist textbooks and 'magnetism' and 'light' together accounted for about another 20 percent.

When textbooks alone were considered, it was apparent that a few central topics did not dominate upper grades of Population 1 and 2 — emphasized topics took up only 25 percent of textbooks in Population 1 and less than 10 percent in Population 2. On the other hand, textbooks for the Population 3 physics specialists used about 70 percent of their space to cover six emphasized topics.

Combining sequence data on the flow of science curricula — the grades at which topics were typically covered — with document-based data on the commonly intended topics at the three key grades revealed a picture of considerable cross-national diversity. This again suggests caution in interpreting science attainment results.

There were surprising commonalities across countries in terms of the topics widely intended at key grades and those that dominated the relevant textbooks. A wider look at focus data for all the grades, however, revealed considerably more diversity.

Emphases in Science Curricula (See Chapter 7)
Science topics are not likely to receive the same amount or kind of curricular attention across their total coverage in a country's curriculum. For some topics, work will go on longer than for others; others will be the subject of intense work for various periods of time. Thus, certain topics will receive more intense curricular attention and emphasis than others over time.

Varied methods were used to assess such intense curricular attention based on curriculum data and documents.

One criterion for curricular attention or emphasis was the presence of topics in both curriculum guides and textbooks at key grades. Many topics received relatively high levels of curricular attention in this sense. For example, 'plant, fungi types' and 'animal types' were present in both the curriculum guide and textbooks for the upper grade of Population 1 in many countries. Other emphases were identified in this way.

An alternative criterion for emphasis or intense curricular attention is whether a topic was one of many covered in the same grade. It cannot be assumed that multiple topics covered in the same grade within a country received equal amounts of curricular attention. However, we can assume that covering several topics in some way limits the curricular attention devoted to each. Thus, somewhat less attention was paid to a topic when it was one of many covered at a grade than would be the case if it were one of only a few topics. In this sense, data presented in chapter 5 on how many topics were covered in each grade in each country give some indication of the emphasis or intensity of curricular attention they received. Some countries were seen to have given comparatively more intense curricular attention to a few topics, while others had a more diverse — and presumably less intensely focused — science curriculum.

The curriculum coverage and focus data were also used to construct numerical indicators of curricular attention for each grade and cumulative curricular attention across all the grades within each country. This method revealed a pattern of marked cross-national variation. Many countries showed little or no comparative emphases on particular topics. The few topics that did receive some emphasis were the same previously identified as emphasized and widely covered.

A more fine-grained approach involving percentages of textbook blocks was used to assess the relative emphasis of all topics and to assess cross-national differences in such emphases. The pattern that emerged clearly emphasized the commonly intended topics identified in chapter 6; even among those commonly intended topics, however, there was substantial cross-national variation. Other topics emphasized in some, but not all, countries were also identified.

This material provides a more detailed context for previously presented data on focus topics and emphasized, commonly intended topics. Certainly, the degree of curricular attention suggested by those designations was borne out by the analyses of this chapter. Further, however, the nature of cross-national variations in curricular attention among topics and grades was made clearer. This served to make the few central topics stand out even more.

Performance Expectations and Disciplinary Perspectives (See Chapter 8)

The sampled science curriculum guides and textbooks revealed widely held commonalities in what students were expected to do with the science contents specified.

- At the upper grade of Population 1, common intentions involved 'understanding complex information,' 'using apparatus, equipment, computers,' 'doing routine experimental operations,' and 'gathering data.' The only commonly intended performance expectation (in the curriculum guides and textbooks of 70 percent of the countries) that was also emphasized (in 6 percent of textbook blocks) was 'understanding simple information.'

- The common expectations for the upper grade of Population 2 were more varied, covering a wider range of performance expectations — 'abstracting, deducing scientific principles,' 'constructing and using models,' 'conducting investigations,' and others. However, only two expectations — 'understanding simple information' and 'understanding complex information' were emphasized.

- For the physics specialists of Population 3, performance expectations were widely present in both guides and textbooks, although less than those present in Population 1 and 2. The performance demands of the textbooks were generally less varied than were those of the guides — the opposite of what held true for mathematics specialist textbooks.

These common patterns again emerged against a background of considerable cross-national variation. This was true whether attention was devoted to which performance expectations were used often either in guides or textbooks, which were used for particular science topics, or which science topics were most often associated with major categories of performance expectations. Similar analyses were conducted on relative use of disciplinary perspectives — the third aspect of the TIMSS science framework — but little of consequence was found.

Investigating Shared Visions (See Chapter 9)

In searching for shared curricular visions, it is also important to look for those shared by some, but not all, of the countries. To this end, we looked at the curricular intentions of countries grouped by identified a priori conditions such as geographical region or average national income.

Regional country groups were examined, and the average percentage of textbook coverage for each topic in each region was compared. There were some differences characteristic of some regions, such as remarkable differences in earth science topic coverage at Population 2 and in treatments of topics such as 'interactions of living things' and some physical science topics in both Populations 1 and 2. Similar analyses were performed for countries grouped into four categories of national average income used by the World Bank — high, upper middle, lower middle, and low. With a few striking exceptions, most of the differences found were small. Why these differences exist is a matter for conjecture and future investigation. A further analysis of European Union (EU) versus non-EU countries revealed few differences and none that were remarkable.

Statistical clustering methods were also used to form groups rather than use a priori categories. In this analysis, we focused somewhat on unexpected similarities revealed, but considerably more on determining which science topics were the basis for the different groups formed. A few topics were important in distinguishing groups. At the upper grade of Population 1, topics such as 'electricity,' 'types of forces,' 'land forms,' and 'bodies of water' were central in distinguishing several groups. 'Electricity' was also important at the upper grade of Population 2, as were 'light,' 'magnetism,' and 'earth in the solar system,' among others.

The results represent only a first pass at analyses for limited country groups. This area seems a valuable topic for further work, both to identify salient differences and to confirm or disprove traditional conjectures about cross-national differences based on such criteria as regional differences or economic affiliations.

Concluding Remarks (See Chapter 10)

One major theme that arose during our analyses was that of pervasive and detailed cross-national variation. Curricular data vary among countries in myriad ways both small and large. Any use of these data must keep this variation in mind. This suggests a second major theme — that any comparative data must be understood in context. Cross-national comparisons are intended to provide information about curricula, educational systems, national goals, and how well national goals are being met. Comparisons can do that only if they are made on factors sensitive to the effects of curriculum and educational systems, and if they are relevant to national goals in science education. The analyses presented here suggest that comparisons will necessarily have to consider sensitive factors — and cross-national differences regarding them — in providing a context for findings and an appropriate basis for interpretation.

Other themes worth noting include possible limitations on a priori group comparisons. It is common in comparative education research to find that groups of nations formed on economic and regional criteria differ in some educationally important ways. The results here, while preliminary, caution against over-optimism that this observation can be validated empirically in the case of curricular factors.

A final pair of conclusions deal with document analysis. Large-scale cross-national comparisons previously have relied mostly on survey and expert data. Document-based data have been reserved for smaller scale studies with limited comparisons. The present investigation suggests that document-based data gathering and analyses are feasible on a large scale and can yield useful data. On the other hand, these investigations have shown how demanding such research is and revealed some of the limitations inherent in the sort of statistical abstracts of documents that appear to be necessary for large-scale document analysis.

More must certainly be done with TIMSS data and more is planned. Investigations are planned for the structure of textbooks and for the relation of textbook form and function. As the full sets of data become available, analyses will be made to relate intention, implementation, and educational attainments. Further studies are needed using these data to investigate intracountry variations in nations with regional and other subnational education systems and varying science curricula. Also, there should be further investigation of limited country groups to find substantive similarities related to national and educational system factors.

The TIMSS ambitious data collection holds promise for future analyses. However, it also revealed important omissions in its design. To rectify the limitations caused by these omissions, direct data are needed on sequence and topic emphasis; documents other than curriculum guides and textbooks must be analyzed; data on the role of textbooks and cross-national differences in that role need to be gathered; and more detailed information on curriculum guides ought to be obtained.

One set of curriculum analysis methods has been used for our large data collection. Other methods can provide additional insights. What we have presented in this monograph is only a beginning. More is planned for primary analyses in TIMSS; more should result from eventual secondary analyses and from the further data collections suggested by these analyses.

II. Curricular Organization and Control

Chapter 3

SEEKING SHARED VISIONS AMONG THE MANY

Our investigation seeks commonalities in the educational visions and aims of the various countries participating in the Third International Mathematics and Science Study (TIMSS). Before those commonalities can be sought, however, we must first acknowledge and understand the differences and similarities across TIMSS countries. These countries differed markedly in their educational systems. For example, they differed in terms of the age at which students entered school, the number of years schooling was provided, the number of years of compulsory schooling, and how grades were clustered. They differed too in the level at which decisions are made, in how complex the internal structures of their educational systems were, and in which decisions are made at each level. And, while countries shared certain broad-based agreements — for instance, virtually all countries required science instruction and tended to cover the same general topics — their approaches to science education varied widely. They structured and used their curriculum guides and textbooks differently; they sequenced their content differently; they had different coverage of science content areas; and they held different expectations of student performance, both globally and within any given topic.

This chapter elucidates some of the key differences and basic similarities across TIMSS countries. This information provides a context against which to understand and evaluate the more detailed curriculum-specific findings outlined in section III.

Comparing the Structures of Educational Systems

National educational systems differ widely in their organization, their curricular organization, their degree of homogeneity, and their decision-making processes. This section examines these diversities across TIMSS educational systems. Only after these essential differences have been distinguished can similarities — broad or specific — in science curricula and educational opportunities be investigated.

System Organization

National educational systems determine school entry ages, the number of years schooling is provided, the number of years schooling is compulsory, and which grades are grouped together.

Table 3.1 summarizes information about these factors gathered from TIMSS questionnaires completed by representatives of participating countries and from related United Nations Educational, Social, and Cultural Organization (UNESCO) data.

Though it remains a matter of some debate among science education researchers, students' cognitive development and maturation are commonly believed to affect, and perhaps even set limits on, what are considered to be curricularly appropriate educational opportunities to foster science learning. The possibilities for providing these educational opportunities differ considerably among TIMSS countries. In most TIMSS countries, pupils began formal schooling at either age 6 or 7. Only three countries — Cyprus, the Netherlands, and Scotland — began compulsory formal schooling before age 6. In three countries (the Russian Federation, Sweden, and Switzerland), students began at different ages (either 6 or 7). There also were differences in the number of years for which schooling was compulsory. This ranged from 12 years in two countries (Belgium and Germany) to 5 years in Colombia. About half of the countries required children to attend school for either 9 or 10 years.

There was more similarity in the total number of years children would have to remain in school to complete primary and secondary education — that is, in the total years of schooling provided by TIMSS countries. The majority required 12 years to complete schooling through upper secondary, and another 13 countries required 13 years. Only three countries required fewer years of pretertiary schooling: Colombia (11), the People's Republic of China (10), and the Russian Federation (10 or 11).

Schooling typically is segmented into two or three clusters of adjacent grades — primary or elementary school, lower secondary school, and upper secondary school. These groupings can reflect differing schooling approaches, policies, resource distributions, student peers, educational opportunity distributions, and other differentiations in schooling possibilities.

Groupings differed considerably among TIMSS countries. More than a third of the countries had about six or seven grades of primary schooling, followed by two or three grades of lower secondary and 3 or 4 years of upper secondary. There were many deviations from this dominant pattern. In Hungary, for example, students followed 8 years of primary schooling with 4 years of secondary.

These differences in the age at which students begin school, the years of schooling, and the organization of grades have important consequences for understanding the curriculum analysis data. There was no common age, for example, for children in grade 7. Countries also differed in whether they considered that grade part of primary or secondary education. Table 3.2 shows how five typical TIMSS countries organized grades in their educational systems. Even this small sample illustrates the considerable organizational differences among countries.

Table 3.2 also shows how these grades correspond to the populations for which TIMSS collected achievement data. By design, these target years also corresponded to grades from which

Table 3.1
Differences in National Educational Systems.

This table presents information on which grades were clustered together, on the number of years of compulsory education and years of education offered, and on school entry ages in various countries. Cross-national differences are clear.

Country	Grade Sequence	School Entry Age	Years of Compulsory Education[a]	Years of School
Argentina	1-7, 8-13	6	7	13
Australia	1-6(7), 7(8)-12 [or 7(8)-10, 11-12(13)][b]	6[c]	9,10	13
Austria	1-4, 5-8, 9-12	6	9	12
Belgium (Fl)	1-6, 7-8, 9-10, 11-12	6	12	12
Belgium (Fr)	1-6, 7-8, 9-10, 11-12	6	12	12
Bulgaria	1-4, 5-8, 9-12	6	8	12
Canada	1-6, 7-9, 10-12	6	10	12
China, People's Republic of	1-5, 6-7, 8-9, 10-12/1-6, 7-9, 10-12	7	9	10
Colombia	1-5, 6-9, 10-11	6	5	11
Cyprus	1-6, 7-9, 10-12	5.5	9	12
Czech Republic	1-5, 6-9, 10-13(14)	6	9	13/14
Denmark	1-9, 10-12	7	9	12
Dominican Republic	1-6, 7-9, 10-12/1-8, 9-12	7	8	12
France	1-5, 6-9, 10-12	6	10	12
Germany	1-4, 5-10, 11-13	6	12	13
Greece	1-6, 7-9, 10-12	6	9	12
Hong Kong	1-6, 7-11, 12-13	6	9	13
Hungary	1-8, 9-12	6	10	12
Iceland	1-7, 8-10, 11-14	6	10	14
Iran	1-5, 6-8, 9-12	6	8	12
Ireland	1-6, 7-9(10), 11-12[d]	6	9	13 or 14
Israel	1-6, 7-9, 10-12/1-8, 9-12	6	11[e]	12
Italy	1-5, 6-8, 9-13	6	8	13
Japan	1-6, 7-9, 10-12	6	9	12
Korea	1-6, 7-9, 10-12	6	6	12

a For some countries the number of years of compulsory education is given only for those students in a general course of study and excludes, for example, those students in vocational programs.

b In Australia, only the state of Tasmania has a substantial number of students attending grade 13.

c In Australia, although compulsory schooling begins at age 6 (grade 1), most children enter a non-compulsory preparatory year with a formal curriculum.

d In Ireland, grade 10 is an optional "transition" year currently being introduced in all schools. Nearly all children enter school at age 4 or 5, attending non-compulsory pre-primary grades with a formal curriculum.

e In Israel, compulsory education consists of 1 year of pre-primary and 10 additional grades.

Table 3.1 *(continued)*
Differences in National Educational Systems.

Country	Grade Sequence	School Entry Age	Years of Compulsory Education[a]	Years of School
Latvia	1-4, 5-9, 10-12	7	9	12
Lithuania	1-4, 5-9, 10-13	7	8	12
Mexico	1-6, 7-9, 10-12	6	9	12
Netherlands	1-8, 9-11, 12-14	4	11	14
New Zealand	1-8, 9-13	6[f]	10	13
Norway	1-6, 7-9, 10-12	7	9	12
Philippines	1-6, 7-10	7(6)[g]	6	10
Portugal	1-6, 7-9, 10-12	6	6	12
Romania	1-4, 5-8, 9-12	6	10	12
Russian Federation	1-4 or 1-3, 5-9, 10-11	6 and 7	8 or 9	10 or 11
Scotland	1-7, 8-13	5	11	13
Singapore[h]	1-6, 7-10, 11-13	6		13
Slovak Republic	1-8, 9-12(13)	6	9	12(13)
Slovenia	1-4, 5-8, 9-12	7	8	12
South Africa	1-7, 8-10, 11-12	7	10	12
Spain	1-8, 9-11, 12	6	10	12
Sweden	1-3, 4-6, 7-9, 10-12[i]	6 or 7[j]	9	12
Switzerland	1-6, 7-9, 10-13 or 1-4, 5-9, 10-12/13	6 and 7	8 or 9	13
Thailand	1-6, 7-9, 10-12	6	6	12
Tunisia	1-6, 7-9, 10-13	6	11	13
USA	1-6, 7-8, 9-12	6	11	12

f In New Zealand, most children begin schooling at age 5, although age 6 is the age at which compulsory education begins.

g In Philippines, school entry age became 6 in 1995.

h General education (Primary 1 to Primary 6 and Secondary 1 to Secondary 4) is provided for every pupil in Singapore. It is, however, not compulsory.

i In Sweden, the grade structure has changed to 1 through 9 and 10 through 12.

j In Sweden, students enter school at age 6 or 7 at parent's choice.

Table 3.2

Organizational Differences in Five Typical Countries.

This display of how grades were grouped in phases of schooling in five countries clearly illustrates the strong organizational differences among countries. Indicating key grades for TIMSS achievement testing makes it clear that these difference have implications for testing.

Country	Primary								Lower Secondary					Upper Secondary			
Hungary	1	2	3	4	5	6	7	8						9	10	11	12
Mexico	1	2	3	4	5	6			7	8	9				10	11	12
Norway		1	2	3	4	5	6			7	8	9			10	11	12
Russian Federation	1	1/2	2/3	3/4					5	6	7	8	9		10	11	
USA	1	2	3	4	5	6			7	8				9	10	11	12

Note: These numbers represent international "grade equivalents" — that is, years in school counted from the beginning year for each country.

Legend: ☐ TIMSS Population 1 Upper Grade   ☐ TIMSS Population 2 Upper Grade   ☐ TIMSS Population 3 Specialists

more detailed curriculum analysis data were gathered. As shown, for example, the upper grade for TIMSS Population 1 for Norwegian and some Russian Federation children was grade 3 rather than grade 4, as was the case for most countries. Similar differences occurred with Populations 2 and 3. If achievement attainments are primarily linked to age-related development, these differences should have few consequences. However, if attainments are related to school arrangements and opportunities, such differences could result in distinct attainment differences.

Curricular Organization

Across the TIMSS countries, the science curriculum is organized differently, with notable variations in the organization and sequencing of opportunities to learn science during specific grades. In many countries, each student received more than one science course per year, and there were a variety of curricular streams in which these courses were situated — especially at grades 10 and 11. For example, students might receive science as a part of secondary science or technology tracks or in a baccalauréate[1] program. Seven major curricular tracks existed in France; no such differentiation existed in Japan. Other data indicate that in Israel, for example, even though there is no formal tracking distinction from grades 1 through 9, students received science instruction not only in their elementary and lower secondary science courses, but in grades 4 through 12 they received earth science instruction in a separate geology/geography course with biology, physics, and chemistry being taught separately from grade 10 onward. Chemistry and physics are also taught separately starting in grade 10 in Mexico, with separate courses on scientific method and biology covered in grade 11. In grade 12 there are four separate courses: biology, chemistry, physics, and geography (which contains earth science topics). But by no means do countries teach the sciences as separate disciplines only in upper secondary. In the Russian Federation, for example, students receive separate courses in biology (grades 6 through 11) and earth sciences (grades 5 through 10), with courses in physics and chemistry beginning in grades 7 and 8, respectively, finishing in grade 11.

Curricular Decision Making

National education systems also varied in their internal complexity. Some were truly single national systems; others were aggregations of smaller (regional, etc.) educational systems. In most cases, an overall national system coordinated the activities of the sub-systems. In other cases, (for example, the United States) the regional educational systems were relatively autonomous with only indirect national influence. Countries ranged from those having a single national system (for example, in Korea) to a complex, varied hierarchy of sub-systems (of which the 50 major sub-systems in the United States were an extreme example). Most TIMSS countries had fewer than five sub-systems.

Educational decision making and policy setting are likely to be affected by the complexity of national education systems. Countries with a single, homogeneous system can conceptualize, articulate, and decide upon curricular changes with relatively greater ease than countries with more complex, heterogeneous systems. Curricular decision making and change become more difficult with increasing heterogeneity.

Spain is a good example of a relatively heterogeneous national system. It had several major types of sub-systems. There were public, religious, and non-sectarian private sponsorship of general basic education, and baccalaureate and vocational sub-systems operated throughout the country. The same types of institutions had sponsored similar sub-systems limited to autonomous communities within Spain. The resulting "national" educational system was one with considerable diversity in curricular authority and in the locus of authority for educational decisions.

Countries also varied in terms of how they approached educational decision making. Decision-making authority may be relatively centralized or more widely distributed, with authority shared by central ministries, schools, teachers, or combinations of these. Within the country's national systems and sub-systems, different agencies had decision-making jurisdiction over different aspects of science curricula. Italy, for example, had only one national educational system but three different sites for decision making: the national ministry of education which had sole responsibility for all decisions relating to national goals, instructional content, and examinations; teachers committees, which were responsible for textbook selection; and individual teachers, who had exclusive responsibility for lesson planning and instructional methods (although the ministry provided advisory recommendations).

Japan, on the other hand, had three educational sub-systems, yet all decisions regarding national instructional goals and content were the responsibility of the national Ministry of Education, with the advice of the Central Council for Education and a sub-committee composed of scholars, teachers, administrators, and persons from industry. Teachers' associations served in advisory roles. No other agencies contributed to decisions on these issues. Prefectural/municipal boards articulated their standards based on the national course of study. Teachers in all three sub-systems decided on instructional technique. Individual upper secondary schools selected the textbooks to be used. In elementary and lower secondary, textbooks were selected by prefectural/municipal boards of education with the advice of teachers.

In TIMSS countries, curricular decisions on national goals, instructional content, examinations, and so on were made by groups, agencies, individuals in authority, or some combination of these. Each country participating in TIMSS was surveyed regarding these decisions and the various authorities involved. Overall, central authorities made more than 40 percent of all curricular decisions, individual teachers about 17 percent of all decisions, and schools about 6 percent. About 26 percent of all decisions were made jointly by more than one source of decision making. Other data show that decisions concerning textbooks were made jointly in about 42 percent of the systems and sub-systems, most often involving a central authority and the school using the text. Individual teachers chose the textbooks for their classes in only about 9 percent of the systems and sub-systems but were responsible for planning lessons in 65 percent of cases.

Table 3.3

Summary of Science Reform Interests in a Representative Set of Countries.

Widespread interest was expressed in science education reforms. This table summarizes reform interests in a representative sample of TIMSS countries as provided by experts within each country.

Country	Remarks
Australia	There is a greater emphasis on the relevance of science and the applications of science in everyday life. The human aspects of science and the impact of science on the environment and society are an integral part of new syllabuses. There is a corresponding change in the methodology of science teaching and learning related to practice and introduction of concepts linked to students' experiences and real world contexts. Programmable and graphic calculators, computers, and interface devices are used for measurement, analysis, and presentation of data.
Belgium (Flemish- & French-speaking)	An important innovation is that much has been accomplished in terms of 'quality control' in Belgian education. Final objectives have been developed for science. These set the minimum standards for what has to be learned. Selected issues include: (1) More attention will be given to the learning of concepts. Graphical solutions will be stressed and more audiovisual means will be introduced; (2) A less encyclopedic approach to the teaching of physics will be stressed; (3) In biology, a more experimental and practical approach will predominate; the program with provide an introduction to an ecological view of the living world; (4) Minimum equipment facilities for science classes will be stipulated; (5) Although the use of computer applications in the diverse fields of science is still limited, attention will be given to technological developments in this area.
Bulgaria	There is some experimentation to integrate science in middle schools, introduce environmental studies, and differentiate teaching according to student interests and abilities. New trends include integrated science in secondary schools and the reduction of material covered. The use of calculators and computers in schools has increased in the past 10 years and software has been introduced for the teaching of chemistry, physics, and biology.
China, People's Republic of	Additions to the curriculum include energy resources, environment, population issues, pollution, electronic technology, computers, and space technology. New technologies in use include microcomputers, VCRs, TVs, projecting apparatus, etc.
Czech Republic	Curriculum contents have been gradually reduced while preserving the concepts that enable a broader region of facts to be described. There is increased attention paid to the effects of technical processes on the environment, both locally and globally, but this has had little effect on curricula. The daily use of calculators, computers, and video equipment is recommended.
Denmark	The following is based on 1988 guidelines. *Physics:* Students must be able to interpret phenomena in our immediate surroundings from a physics point of view; must acquire the idea of physics as a coherent description of nature; must learn the history and theory of physics and its applications; must acquire an insight into the close relationships between progress in physics and the development of society and technology. Use of calculators in physics instruction has been permitted since 1976 but is not allowed at written examinations. *Chemistry:* Students must learn the methods and applications of chemistry in everyday life; must be acquainted with publications from environmental organizations. Computers are used for collecting and processing chemical test results as well as for searching external data bases, writing, and calculating. *Biology:* There has been a shift of focus to analytical problem-oriented teaching as

well as new theoretical and practical knowledge and research. New topics include disease prevention, biological engineering, medical technology, production technology, etc.

A new act affecting the Folkeskole came into force in 1994. The new curriculum guides set new goals and objectives for all science subjects, and are structurally different from the ones used in the present TIMSS study. Additionally, a new science subject has been introduced in grades 1 through 6: "Nature and Technology." Furthermore, it is stressed that "a green weft" is to be in all subjects in the Folkeskole, not only science.

France | In recent years, especially in biology/geology, more importance is given to the teaching through problems around concepts. The general approach is more naturalist and less conceptual. There has been an increased emphasis on technical aspects; on science's everyday applications; and a more systematic approach to the study of life processes (e.g., human nutrition). In physics, changes are motivated by the desire to prepare pupils to interact efficiently with a technological world. Since 1992, physics and chemistry begin at grade 8 instead of grade 6. There is a concern to develop a critical mind, integrate new technologies and computers, and, consequently, individualize learning. There has been an introduction of computer science and computer assisted experimentation in both biology and physics. Software is making the learning of some theoretical concepts easier, allowing the teaching of physics in some additional fields. New programs have been or will soon be written for grades 8 through 12.

Japan | National courses of study were revised in 1989 and are not expected to be revised for another 10 years. The biggest change in the last 10 years is a shortening of the number of hours for science. The relationship between study and daily life is gaining greater importance at all levels. In elementary schools, science is included in "life environmental studies," which was just added to the curriculum. In lower secondary, elective science courses were added and the use of computers was introduced. Students study about the computer as a tool of information science, the computer's development, and development of computer chips. In upper secondary, there has been a tendency to make contents of science easier.

Russian Federation | Changes during the past 10 years did not involve curriculum structure. The change has been from mastering knowledge and skills to developing learning skills. Environmental issues have been given greater attention. The introduction of 4 years of primary schooling (instead of 3) in 1988 did cause structural changes in the elementary science curriculum - two new courses were added: "Surrounding World" for grades 1 and 2 and a "Nature Study" for grades 3 and 4. Due to social changes the entire system of public education is more democratic and humanistic. Two divergent science programs have been proposed for the middle school: the study of all science subjects separately in grades 5 through 9; an integrated science program in grades 5 and 7 with differentiation in grades 8 and 9. In upper secondary school, a basic level of science courses is provided and advanced courses are offered for students who choose programs in physics-mathematics, science, or technical studies. A new course "Informatics" has been added to the curriculum and schools are being supplied with computers. However, most schools don't have enough computers and calculators so there are no concrete recommendations for their use in science instruction. Materials describe general trends in technology in connection with societal problems, primarily because technical descriptions change so quickly.

USA | Historically the rationale for science instruction was to prepare a sufficient number of scientists and engineers for the nation. Since 1983 science instructors have been urged to prepare citizens for productive lives in a society that increasingly is dependent on science and technology. As a result there is increased emphasis on earlier and more science instruction in elementary schools. There is an increasing concern with environmental issues and increased attention to the interactions among science, technology, and society. There is movement away from many topics to more in-depth instruction on a limited number of topics. One way to accomplish this is "themes" that are applicable across all science fields. Another change is moving from instruction of "facts" to hands-on doing of science. Testing is affected by this development. There is increased emphasis on using real-world settings and involving students in problem solving. A very significant reform is the development of national standards for science curricula, assessments, and teaching by the National Academy of Sciences. Technology exists increasingly in schools, but it tends to be supplemental or peripheral rather than integrated. For schools that do have technology, students are able to use telecommunications to exchange scientific data and to access scientific databases.

Broad Similarities in Science Education

While differences clearly existed in the educational system structures and practices of TIMSS countries, it is equally clear — and equally important — that broad similarities were also present across countries. Out of the possible similarities, we focus on two here.

Required Science Instruction

All the TIMSS countries required some science instruction in both primary and secondary school. The amount of science covered at any given grade varied, especially in the first elementary grades, but the constant across all countries was that some science was always required. Some countries focused on fewer topics at a time, others on more. Using the most detailed categories from the TIMSS science framework as a list of possible topics, the number of science topics intended to be introduced by the end of secondary school ranged from 38 to 78.

Beginning in lower secondary school, the situation is quite complex when considering the number and type of courses in which these topics are covered. Although every country participating requires science at this level, approximately 60 percent require more than one course — each in a separate scientific discipline or set of disciplines. Thus, some countries require a course in biology and another in physics. Others have a combined biology/earth sciences course and a course in physical sciences that combines both chemistry and physics. Still others intend science to be taught in courses called *orienteering* and *anthropology*, or even present considerable amount of earth sciences topics in geography, history, or social studies courses. The remaining countries require an integrated science course, but the disciplines represented in the content of these courses can vary substantially from country to country.

Reform in Science Education

Among the TIMSS countries, there was a widespread interest in reform of science education: many reforms and managed changes were taking place in science content, pedagogy, and technology use. TIMSS surveyed panels of experts in each participating country concerning ongoing reforms in their countries — both recent and projected near-term reforms. Table 3.3 summarizes these comments on science education reform initiatives for selected countries. (This information may be found in table I.3 in appendix I for all countries that provided such data.)

Common themes emerged among these reforms. One was the importance of teaching science in the context of everyday applications. Common justifications argued for deeper student engagement and greater utility of the science taught. Denmark, France, Japan, the United States, and Australia were among the countries reporting reform efforts oriented toward applications and meaningful real-world activities. How extensively these goals had been implemented at the time of the survey varied greatly. Some countries even looked to TIMSS for information on how far such reform had progressed.

New technologies were influencing thinking about science curricula. Many TIMSS countries reported encouraging calculator use in teaching and learning science in secondary schools. Australia, the Czech Republic, and the United States reported such reforms; some countries reported beginning to permit calculator use on national science examinations. Science curricula were also reported to be influenced by recent advances in computers and software.

Other countries reported adding topics or increasing emphasis on ecology, pollution, and the environment. The People's Republic of China, Japan, the Russian Federation, Belgium (both systems), and others reported such changes in the natural sciences curriculum. Some countries, such as Bulgaria, reported experimentation with teaching integrated science in lower secondary school, moving away from a tradition of separate courses in the sciences.

Notably, only one country, Japan, reported a *reduction* in the hours devoted to science instruction. This seems to be a reform that contradicts trends prevailing in most other countries participating in TIMSS.

Differences in the Details of Science Education

Organization and Numbers of Science Courses

The order and number of science courses offered in TIMSS countries varied substantially. Although there was considerable variation, almost 60 percent of the TIMSS countries offered more than one science course in the upper grade of Population 2, and 51 percent offered three or more courses at this grade.

Differences in how science was taught in grade 10 in New Zealand, Colombia, and Hungary illustrate the variation across countries. Grade 10 was the second year of secondary school in New Zealand and the modal age of students was 14. Grade 10 was the first year of the upper secondary sequence in Colombia and the modal age of students was 15. In Hungary, students in grade 10 were in their second year of upper secondary school and the modal age also was 15.

Only one science course was offered in grade 10 in New Zealand. This general science course covered a variety of topics in the physical and natural sciences; all students at this grade level were enrolled. The number and length of periods for which this course met varied from school to school. As noted earlier, less than half of the TIMSS countries offered only one integrated science course at this grade level.

In Colombia, grade 10 was the first in which the sciences were taught in different courses according to discipline. One course was offered in physics, another in chemistry. Students were required to take both courses, and 100 percent of students at this grade level are enrolled in each. The physics course covered mostly topics in energy and physical processes as well as related topics in technology. The chemistry course covered topics regarding matter and its physical and chemical properties, structure, and chemical transformations. Both of these courses met for two to four 60-minute periods per week.

Hungary has several grade 10 science courses. All grade 10 students were in the second year of their 2-year secondary geography and chemistry courses, both of which met for about three 45-minute periods per week. They also were in their second year of a 4-year course entitled "Secondary Physics" that also met for two to three 45-minute periods a week. In addition, 50 percent of these students were enrolled in at least one 3-year optional course in either biology, physics, geography, or chemistry, which met for three to four periods a week.

These courses met for different amounts of time the content varied, and course goals varied. Students, instructional resources, and teaching personnel were organized differently. Obviously, science meant different things to grade 10 students in these three countries.

This variety of courses in many countries — each often with its own textbooks and guides — presented a challenge to the collection of data on the science curriculum. All courses intending instruction of topics in the TIMSS science frameworks were included in all data collections on topic coverage, textbooks and guides. For some countries this meant the materials for courses with titles such as *geography, anthropology, history of the fatherland,* or *orienteering* were included.

Curriculum Guides and their Use

Virtually all educational systems use some form of curriculum guide to structure science education. These guides set forth the system's goal for its science education. Countries differ widely in the structure and details of their guides and in how they are meant to be used.

Curriculum guides serve many purposes. They may specify aims and objectives, set out policies, identify content to be covered, suggest teaching or testing strategies, or present examples to guide teachers. One broad measure for gauging the various purposes of countries' curriculum guides is page length. Among TIMSS countries, the page length for curriculum guides for the upper grade of Population 2 varied from less than 10 pages to hundreds of pages. Even after adjusting for differing page dimensions, this is still a significant variation. It indicates differences in guides' level of detail, and likely reflects differences in intended use.

A more refined indicator for identifying curriculum guide purpose is to determine the percentage of each guide devoted to shaping specific aspects of the science curriculum. Each text block of each guide was coded to indicate whether it addressed an aspect of the TIMSS science framework. Blocks that received no codes were probably those that made general suggestions or set out policies. This approach compensates somewhat for the extensive variation in the sizes of sampled curriculum guides — block *percentages* are relatively stable and characteristic of the guide, regardless of numbers of blocks contained.

More than half of the guides had one framework code (see explanation in appendix C) for more than 90 percent of their blocks. Only two countries had less than 70 percent: Iceland and Scotland. This high level of content specification seems to have been independent of the length of the guides.

Textbooks and Their Use

Pedagogical strategies and approaches vary among countries even when similar science content is being presented. That point is driven home by examining textbook data: the lengths vary, the sizes vary, the formats vary, and the levels of decoration vary. A scatterplot of textbook size (that is, page length times width in square centimeters) versus the length (number of pages) would show that textbooks came in virtually every combination. The one comparatively rare pattern was small books with few pages. The largest books had more than three times the page area of the smallest. The longest were six and seven times the length of the shortest. Overall, variation was considerable and almost random.

Some countries, notably the United States and Canada, placed enormous books in students' hands. Other countries, such as Scotland and Iceland, offered their students relatively slim volumes. Such extensive variation suggests that the books probably were meant to be used in different ways. A small or slim book might contain only exercises or exposition, but likely was not meant to be used independently of the teacher. On the other hand, a larger text probably contains a greater range of materials (both exposition and exercises) and may have played a larger role in classroom instruction.

One way to delineate differences in book structure and emphasis is to look at the percentages of different block types. These block types, as discussed in chapter 1, include narrative (exposition and explanation), graphics, exercises, activities, and worked examples.[2] If one country's science textbooks contained a comparatively higher proportion of one block type — for example, activities — than did textbooks from another country, it is likely that different approaches were used in presenting content in the textbooks, and possibly textbooks were used in different ways in the classrooms. (It is important to note that countries were required to analyze any separate laboratory manuals that were a required part of science instruction, in addition to analyzing textbooks.)

Figure 3.1 presents the percentages of major block types within the science textbooks for the upper grade of Population 2 for all TIMSS countries. Countries clearly varied in the ways in which they structured science textbooks. Some depended more heavily on narration, with or without graphics (for example, French- and Flemish-speaking Belgium, Iran, and Italy). Some textbooks made at least moderate use of activities (for example, Germany and Denmark), and even more made use of graphical material (for example, Portugal, Iran, and the Russian Federation).

Science Content Areas and Performance Expectations

Each national education system pursued broad topic areas of science that reflected their national vision and aims for science education. These areas and emphases differed among countries — even at the broadest content category level.

Figure 3.2 presents for the upper grade of Population 2 in selected TIMSS countries the percentage of science textbook blocks allocated to the eight broadest content categories from the

Figure 3.1
Distribution of Block Types in Science Textbooks.
The percentages of selected block types are given for science textbooks for the upper grade of Population 2.
The textbooks certainly differed in what they presented and, consequently, in how they might be used.

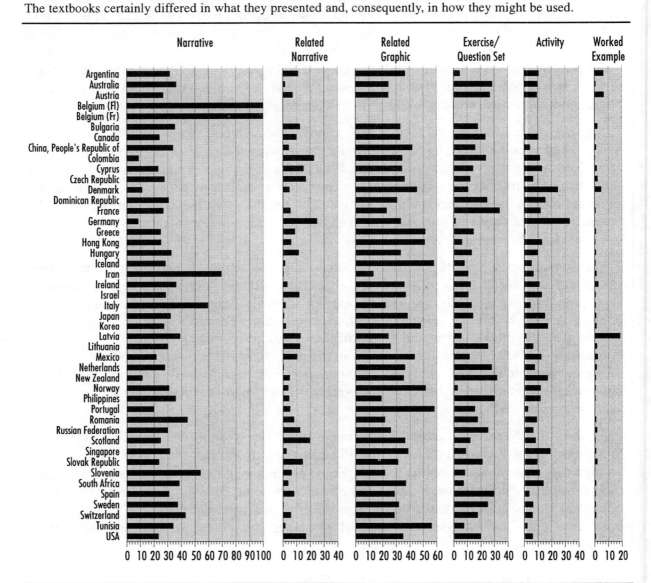

Figure 3.2

Textbook Space Devoted to Major Science Topics.

This figure shows the percentages of textbook blocks devoted to the eight broadest science framework content categories for the textbooks of the upper grade of Population 2 in a representative selection of countries. This shows patterns of varying emphasis that held true for other countries and grades.

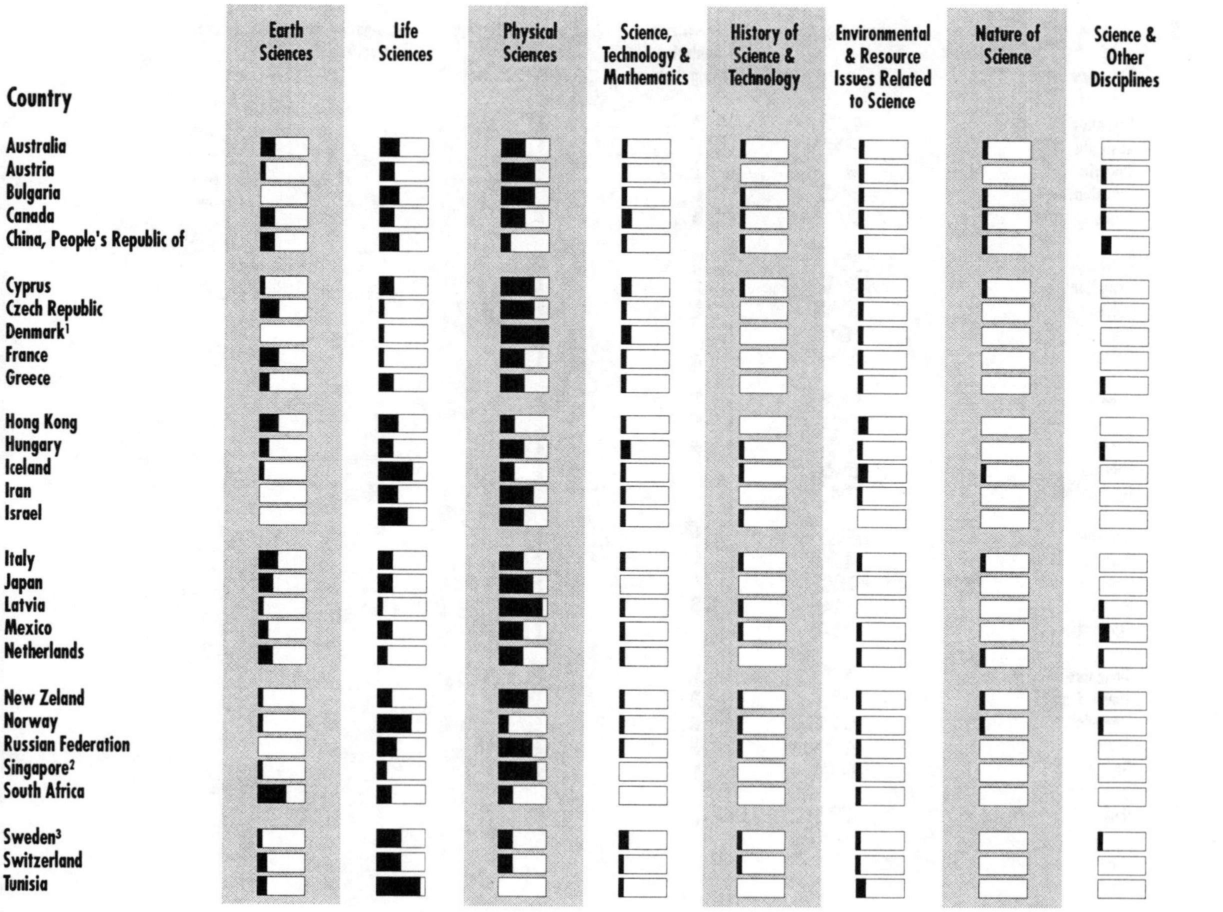

Country	Earth Sciences	Life Sciences	Physical Sciences	Science, Technology & Mathematics	History of Science & Technology	Environmental & Resource Issues Related to Science	Nature of Science	Science & Other Disciplines
Australia								
Austria								
Bulgaria								
Canada								
China, People's Republic of								
Cyprus								
Czech Republic								
Denmark[1]								
France								
Greece								
Hong Kong								
Hungary								
Iceland								
Iran								
Israel								
Italy								
Japan								
Latvia								
Mexico								
Netherlands								
New Zeland								
Norway								
Russian Federation								
Singapore[2]								
South Africa								
Sweden[3]								
Switzerland								
Tunisia								

[1] Denmark only provided data collected from physical science textbooks for this population.

[2] The Curriculum Guide for Singapore was revised in 1992. The National Research Coordinator estimates that in the revised textbook the percentages devoted to 'Physical Sciences' and 'Life Sciences' have changed to 64% and 30% respectively.

[3] Although chemistry books do exist at this level, none were coded.

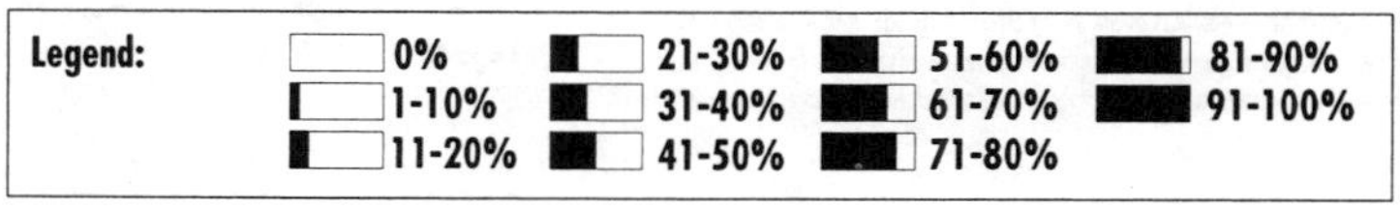

Legend:

0%	21-30%	51-60%	81-90%
1-10%	31-40%	61-70%	91-100%
11-20%	41-50%	71-80%	

Figure 3.3
Textbook Space Devoted to Major Performance Expectations for Selected Countries.
These data are similar to those of figure 3.2, but for the broadest performance expectation categories. They also reveal considerable variation.

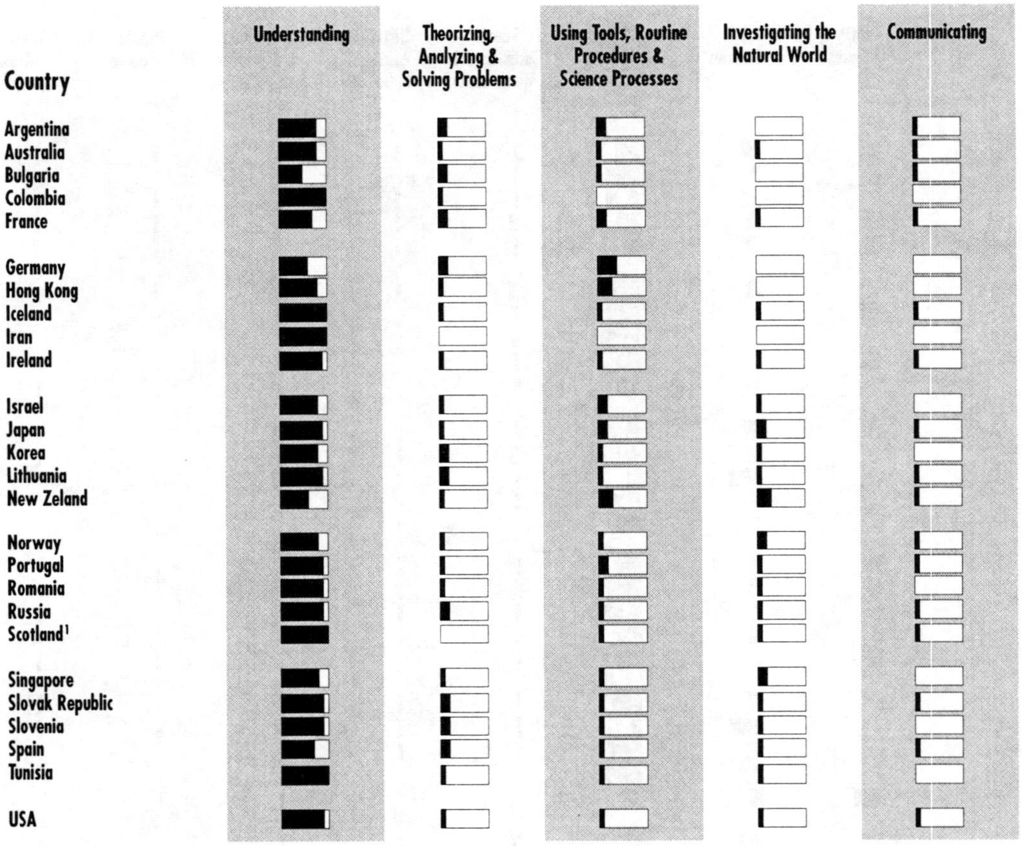

Legend:
0%	21-30%	51-60%	81-90%
1-10%	31-40%	61-70%	91-100%
11-20%	41-50%	71-80%	

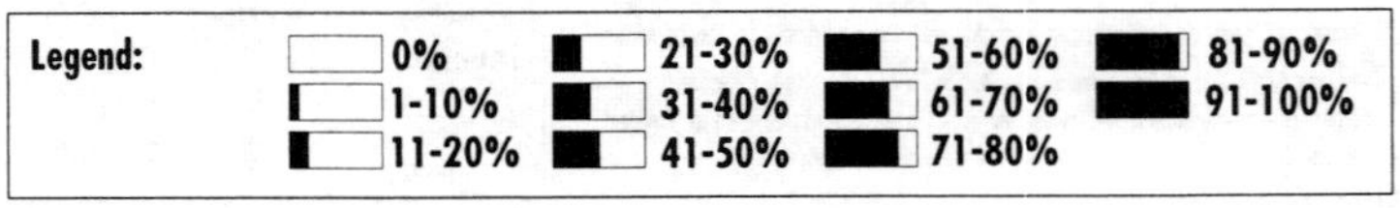

TIMSS science framework (see appendix I, figure I3.2 for all countries). These data show the relative emphases on different science content topics as indicated by space devoted to each.[3]

Figure 3.2 reveals considerable differences in the emphasis on various broad science topics. Some books devoted half or more of their blocks to 'life sciences' (for example, Iceland and Tunisia), while others devoted less than 10 percent (for example, Latvia and Denmark devoted 3 percent). As the figure shows, almost all topics received considerably varying emphases among the countries. 'Earth sciences' ranged from 0 percent (Iran, Israel, the Russian Federation, and Bulgaria) to 51 to 60 percent (South Africa); 'physical sciences' ranged from 0 percent (Tunisia) to between 81 to 90 percent (Latvia). These data can also be used to indicate the topics emphasized by a country's textbooks. For example, Iceland emphasized 'life sciences' over 'physical sciences' and 'earth sciences,' whereas other countries have textbooks that divide their coverage quite evenly among earth, life, and physical sciences (such as Hong Kong). Another pattern was one of divided emphasis between the two disciplines of earth and physical sciences with little coverage of life sciences, as is true for the textbooks of the Czech Republic and France. Many countries concentrate mostly on the physical sciences (Austria, Cyprus, Latvia, etc.); less frequent was a concentration on life sciences (Iceland, Norway, Tunisia).

Textbook data can also be examined for their student performance expectations. Figure 3.3 presents, for selected countries (see appendix I, figure I3.3 for all countries), the percentage of the upper grade of the Population 2 science textbook blocks devoted to the highest level performance categories from the science framework. These categories — 'understanding,' 'theorizing, analyzing and solving problems,' 'using tools, routine procedures and science processes,' 'investigating the natural world,' and 'communicating' — represent major types of performances that might be expected from students in learning and doing science. 'Investigating the natural world' expectations ranged from 0 percent (six countries) to 21 to 30 percent (New Zealand). 'Theorizing, analyzing and solving problems' expectations ranged from 0 percent (Iran and Scotland) to some countries that had between 11 and 20 percent (such as Korea and Spain). 'Using tools, routine procedures and science processes' expectations ranged from 31 to 40 percent (Germany) to 0 percent (Colombia). Expectations for 'investigating the natural world' were quite uncommon, with many countries reporting less than 10 percent of blocks. 'Communicating' was the least common expectation, with a large number of countries reporting no blocks with this performance expectation. From this table it is clear that the most common performance expectation is 'understanding,' with some countries (such as Scotland and Iceland) focusing on this expectation almost exclusively. The notable exceptions are countries that, while still emphasizing 'understanding,' have a significant minority of blocks devoted to 'using routine procedures' (such as Germany, New Zealand, and Hong Kong) and 'theorizing, analyzing, and solving problems' (such as Korea and Spain).

Science Topic Sequences

The sequence in which various science topics are presented varied by country. The combined influences of different educational system structures, science content emphases, and performance expectations across topics yielded varied patterns of topic coverage over the years of schooling. Figure 3.4 presents comparative data for two countries — the People's Republic of China and Iceland — across the range of grades for four topics from the science framework. In Iceland, the topic 'planets in the solar system' was introduced when students were age 13 and emphasized when they were age 14. At age 14, the topic of 'beyond the solar system' was introduced and studied until the end of upper secondary school. In the People's Republic of China, both topics receive focused instruction when students are age 16. In the People's Republic of China, both 'variation and inheritance' as well as 'evolution, speciation, diversity' are taught over a four-year period beginning at age 14. In Iceland, more years are devoted to these topics beginning at ages 14 and 13, respectively.

Do Common Visions and Aims Exist?

This chapter has emphasized that educational systems differ markedly in the years in which schooling begins, the years of schooling provided, the years of compulsory schooling, and in how the grades are clustered. Countries differed in the level at which decisions were made, in how complex the internal structure of the educational system was, and in what types of decisions were made at different levels. Broad similarities existed in that virtually all countries required science instruction and that many of the general topics were common among countries. When we examined details, however, we found that curriculum guides varied in structure and use, as did textbooks, content sequences, science content areas, and performance expectations — both globally and within any given topic.

If commonalties existed only at the broadest levels, is there any point in seeking meaningful common visions and aims by which countries can learn from mutual benchmarking and cross-national comparisons? This remains an empirical question — and one which in part this volume is devoted to answering. Clearly, flexibility must exist in seeking the bases for comparisons. Commonalities can be seen in some features but not in others, and how critical these features will be remains to be seen. When commonalities are sought among smaller sets of peer nations, rather than in the full set of TIMSS countries, stronger similarities may be identified. The remainder of this volume is devoted to setting out some of the many, complex commonalties and differences about science curricula examined through the methods sketched earlier and setting the stage of varied practice across the TIMSS countries.

Figure 3.4
Comparative Sequences in Four Science Topics for the People's Republic of China and Iceland.
Focusing only on two countries and four topics shows that the sequence of covering aspects of topics varied as well as did overall emphasis.

People's Republic of China

	6	7	8	9	10	11	Age 12	13	14	15	16	17	18
Planets in the Solar System											+		
Beyond the Solar System											+		
Variation & Inheritance									+	-	-	+	
Evolution, Speciation, Diversity									+	-	-	+	

Iceland

	6	7	8	9	10	11	Age 12	13	14	15	16	17	18
Planets in the Solar System								-	+				
Beyond the Solar System									-	-	-	+	-
Variation & Inheritance									-	-	-	-	+
Evolution, Speciation, Diversity								+	-	-	-	+	+

Legend: - indicates intended coverage at that age.
+ indicates intended focal coverage at that age.

NOTES

1   The 'baccalauréate' is a special diploma awarded in France based on a student's performance on a set of examinations taken in the final year of secondary school. Possession of this diploma is required to be eligible for admission to a university.

2   Detailed explanation of blocks and block types can be found in appendix E.

3   Textbook emphasis data have certain intrinsic weaknesses. For one, they do not necessarily reflect official curricular intentions in some countries. Additionally, many parts of science textbooks may never have been covered; conversely, there may well have been topics emphasized in class that were not reflected by the space accorded in textbooks. Generally, however, content emphases in textbooks tend to indicate the content that is most likely covered.

Chapter 4

CURRICULUM GUIDES: DIRECTING CURRICULUM BY INTENTION

A group sponsored by a ministry of education meets. Their task is to revise, reform, and renew the science curriculum for public lower secondary schools. The group includes ministry representatives, some scientists and science curriculum specialists from universities, and a few science teachers from lower secondary schools. They talk; they review the current curriculum; they argue. In time, a new vision begins to emerge, a goals statement is drafted that makes that vision specific, and plans are drawn for a new science curriculum.

In the same country, a group of teachers meet to plan science instruction in their grade of lower secondary school for the coming year. They talk; they review documents from the education ministry that are to guide the course of study they will provide; they argue. Their plans emerge. They begin to prepare specifically for day-to-day science instruction.

Later a group of science students sit in the classes of one of these teachers. They listen; they think; they work on assigned tasks; they argue with each other. The teacher's plans change in small ways — as always when faced with an actual class — but mostly she acts according to the plans she made earlier and discussed with her colleagues. The vision of science instruction in her head — and in her plans — is played out in dozens of small scenes with the students in her classroom.

What connects these three groups — curriculum planners, science teachers, and students — is a *curriculum guide*.[1] Activities follow from curriculum planners' ideas, they guide teachers' work, and shape students' efforts.

A curriculum guide is, in most countries, the main document for officially setting out a curriculum. Guides come in many forms, but they all serve several important functions. They represent officially sanctioned decisions, set out policies related to a specific curriculum and its implementation, address those who will implement curriculum (school officials, teachers, etc.), and state aims and goals. These documents also may try to convey the vision lying behind the specific goals and requirements of a particular curriculum.

At their most detailed, they contain suggestions and even examples of teaching strategies and, perhaps, testing strategies as well. Often, they contain detailed content specifications and content sequences to be covered. They are the link between curriculum planners' intentions and implementations of science curricula. In fact, curriculum guides are one of the primary means by which the planners communicate with the implementors and have an impact on their actions.

On the other hand, although curriculum guides are important and serve a roughly similar function for national educational systems, they are also quite varied. This chapter uses TIMSS sampled curriculum guides to present some of the common features and variations in these crucial curricular documents.

Authority and Function

Curriculum guides published by national or sub-national governmental bodies existed in all TIMSS countries except Iran. The guides all carried at least some official status, although status and authority varied among countries and often within a country for sub-national or regional guides. Local guides — for example, for a school or a group of schools — were also often available, but such guides typically interpreted national goals stated in other documents, with the aim of ensuring uniformity and clarity in the local implementation of the curriculum.

The perceived significance varied substantially by country. Curriculum guides in Australia, for example, had such titles as "Course Advice," whereas in Japan they were known as "National Courses of Study" and in Ontario, Canada, as "the Common Curriculum." These diverse titles suggest different official statuses and functions. Some guides specified the courses of study for which teachers were responsible. Others specified *how* teachers might pursue their goals, as well as what types of instructional methods and assessment strategies might be appropriate. Still others left most implementation details to teachers and attempted to accomplish their purpose solely by stating shared objectives.

Grade Spans Covered by Curriculum Guides

Some countries had long curriculum guides that covered numerous grades; others had shorter, more specifically targeted guides. If one focuses on a sample of curriculum guides that include students at age 13, the defining age for TIMSS Population 2, to determine the span (number of grades, that can also be expressed as the range of ages) covered by each guide, the curriculum guides that included 13-year-old students would be seen to have ranged from those that covered only one age to those that covered grades for a span of 14 years. There was considerable variation among countries in the span of years covered, and even considerable variation within countries. Eighteen countries had more than one guide sampled for age 13. The differences between the least and most inclusive varied from 0 to 12 (United States) or 9 (Scotland) years difference. Guides also differed in how age 13 was grouped with other ages. In some cases it was considered alone or with a few other ages; in others was placed near the end of a longer span covering lower grades; in still others, it was placed near the beginning of a span covering higher grades; occasionally, it was near the middle of a guide that spanned virtually all years.

A closer look reveals other patterns among the sampled guides. Five of the sampled curriculum guides were specific to age 13, the target age for Population 2. Another 12 guides covered only 2 years. Of those covering 2 years, there were three guides for which age 13 is the upper age of the span and eight guides for which it is the lower. Eight other countries had guides that spanned 3 years — and in five guides, age 13 was the middle of the 3 ages spanned. In three guides, age 13 was at the beginning of the three ages spanned. In another 10 cases, four grades were spanned, with both the guides sampled for the Slovak Republic being of this type. In another four cases, five ages were spanned, and in another three cases, six or seven ages were spanned.

Finally, other guides covered 8 to 14 years, thus spanning most of the ages for formal schooling. In all of these guides, age 13 was one of the *later* ages included — again suggesting this age was grouped with secondary schooling. Although sampling prevents drawing a firm conclusion, it is worth noting that many longer span guides were from either the United States or countries in the British Commonwealth tradition — Australia, Canada, Scotland, and New Zealand.

Structure of Curriculum Guides

From the sampled documents, it appears that science curriculum guides vary in the strategy they take to set forth and shape curriculum. Specifically, some tend to be more *prescriptive* than others: these state policies, formal objectives for science instructions, and so on. Others tend to be more *facilitative*; they include such information as suggested strategies for teachers, examples, and assessment ideas.

Types of Units

The document analysis methodology for curriculum guides has the ability to capture something of this range from prescriptive to facilitative. Curriculum guides were divided into several types of units.

- *Policy units* were limited to those that stated formal policies about science curriculum and instruction, other than specifications of content or curricular objectives.

- *Objective units* were mainly formal statements of objectives for science curriculum and instruction. These units typically included discussion of the science content involved, what students were expected to be able to do when this content was mastered at the level desired, or both.

- *Content units* were limited to stating specific science contents to be covered when contents were not stated as part of the instructional objectives. In some cases, this information included the relative amounts of emphasis or instructional time allocated to those contents.

- *Pedagogical units* included material that made pedagogical suggestions — for example, strategies that might be used to obtain a particular objective or to cover specific science contents. These units also included examples given of pedagogical approaches, demonstration problems, etc., as well as specific information or examples given about the most effective ways to assess mastery of particular science contents and attainment of particular goals. If, however, the material was actually a policy prescribing standards or procedures for assessing and grading student work, it was classified as part of a policy unit rather than as part of a unit of pedagogy.

These unit types — introduction, policy, objective, content, pedagogy, and others — were designed to provide insight into the approach of each guide. The introduction and other units were considered neutral in or irrelevant to characterizing the approach of curriculum guides; the policy, objective, and content units were more directive and prescriptive as ways of guiding curriculum; and the pedagogy units presented a more facilitative approach to curriculum.

Figure 4.1 displays the percentage of pages[2] for each of the curriculum guide unit types for the Population 1 science curriculum guides in the TIMSS countries with the highest percentages for each of the unit types — especially policy, content, objectives, and pedagogical suggestions.

As shown in figure 4.1, content units were present in most Population 1 science curriculum guides. Flemish-speaking Belgium, Sweden, Romania, and the Russian Federation are among the countries with more than 60 percent of their pages given to content units. Pedagogy units formed more than 40 percent of the material in the curriculum guides of French-speaking Belgium and Israel. Policy units were less common — such units were absent from the curriculum guides of more than one-half of the countries. Norway is an important exception, with more than 40 percent of the pages being devoted to policy units. Many countries devoted much larger proportions of their guides to objective units. Spain and Mexico were remarkable in devoting more than 75 percent of the pages in their Population 1 science guides to units expressing instructional objectives.

Policy, objective, and content units represent a prescriptive approach to setting out curriculum. Aggregating these data shows that some countries took this approach almost entirely. For example, more than 70 percent of the science curriculum guide pages were devoted to one or more of these three types of units in several countries — Flemish-speaking Belgium, Hungary, France, Ireland, Japan, Romania, Scotland, Sweden, Spain, and Norway. This suggests that many countries took a more prescriptive approach to communicating curricular intentions and shaping the implementation of Population 1 science curricula. This does not, however, take into account the possible existence of separate pedagogical documents in some countries that were not analyzed.

In contrast, few countries devoted a considerable number of pages to pedagogical suggestions. Israel, however, devoted more than one-half of its curriculum guide to pedagogical suggestion units. A few countries used a strong supplement of pedagogical suggestions to support

Figure 4.1
Proportions of Different Units in Population 1.

The figure shows the percentage of pages in science curriculum guides for the upper grade of Population 1 for each unit type. Some countries were more prescriptive—that is, had higher percentages of policy, objective, and content units. Others were relatively more facilitative—that is, had comparatively higher percentages of pedagogy units.

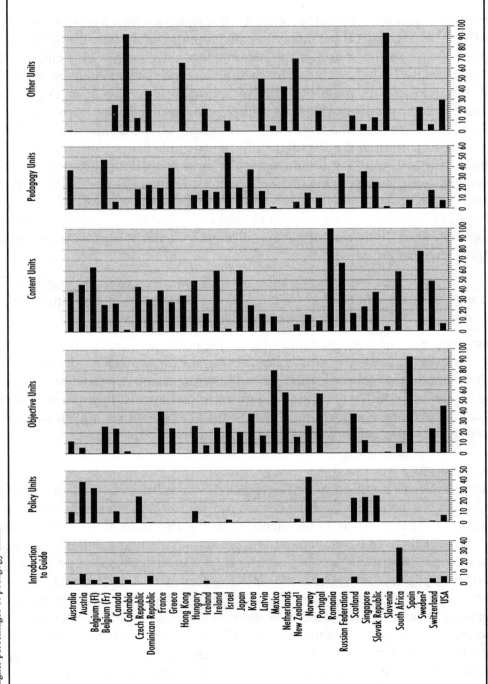

1 New Zealand: TIMSS analyses were based on data from the 1989 revised edition of the curriculum guides that were valid until 1995. New guides were published in 1993 and schools were advised to work toward their implementation at the beginning of 1995.

2 In Sweden, curriculum objectives apply to groups of grades, not individual grades. Caution is suggested when interpreting these data.

   Missing Data: Argentina, Bulgaria, People's Republic of China, Cyprus, Denmark, Germany, Iran (no curriculum guides exist), Italy, Lithuania, Philippines, and Tunisia.

other, more prescriptive unit types. Korea and French-speaking Belgium appear to exemplify use of this combined prescriptive-facilitative approach.

Population 1 data were shown in figure 4.1 because it seemed a reasonable assumption that the most use of pedagogical facilitation (and proportionately less use of more prescriptive approaches) would be found at this population concerning science curricula for the youngest students targeted by TIMSS. Clearly, this assumption was not supported by the data. The same patterns of varied uses of more prescriptive approaches were seen in Population 2 guides. Comparatively less space was given to pedagogical support in the Population 1 materials of many countries. However, in a few cases — for example, the Dominican Republic and Korea — more than 40 percent of the Population 2 science guide pages were devoted to pedagogical suggestions. Flemish-speaking Belgium was the most anomalous case, with no pages devoted to pedagogical suggestions at Population 1 and more than 35 percent at Population 2. In Population 3, curricula — physics for advanced, specialist students — were consistently approached in a less prescriptive way. In several countries, however, such as Spain, Korea and Denmark, more than 40 percent of advanced science curriculum guide pages were devoted to pedagogical suggestions.

It appears likely that Population 1 science curricula were generally treated as comparatively less complex and less in need of formal pedagogical support in curriculum guides than was the more complex and difficult-to-teach material for Population 2.

What is apparent, however, is that a variety of approaches were taken to communicating and shaping curricular intentions in curriculum guides at every curriculum level and that, in most cases, these approaches tended to the more prescriptive and directive in Population 1. The amounts of space devoted to pedagogical support varied interestingly among the countries and the student populations. Clearly this matter of the strategy for setting out science curricula in curriculum guides is a subject worthy of further inquiry.

Type of Blocks

Data are available to take a more fine-grained look at the structure and strategy of presenting curriculum in guides. Because the curriculum guides were not tied to allocated time or any other uniform indicator, we could not identify relative emphases of the guides at the unit level. Instead, in our analysis of curriculum guides, we sub-divided each unit into blocks. Seven different block types were used — policy statement, single objective, single content specifications (element), pedagogical suggestions, example (as part of a suggested pedagogical approach), assessment suggestion, and other. Although these blocks varied considerably in size, and thus in the amount of emphasis indicated, they did provide us with further detail on and insight into the structure and use of curriculum guides. Figure 4.2 displays the variation across TIMSS countries in the use of different types of blocks in the curriculum guides of the upper grade of Population 2, which are similar to that noted for unit types in Population 1.

Figure 4.2

Proportions of Different Block Types in Population 2.

This figure presents science curriculum guide block type percentages.

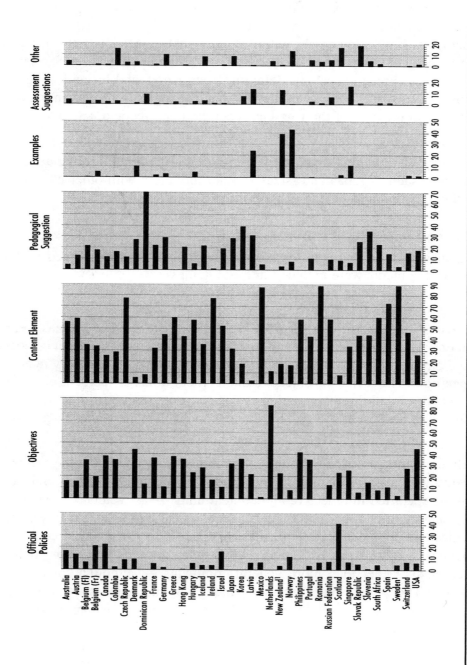

1 Refer to footnotes in figure 4.1.

Missing Data: Bulgaria, People's Republic of China, Cyprus, Iran (no curriculum guides exist), Italy, Lithuania, and Tunisia.

Figure 4.3 demonstrates the variation within pedagogy units. Included in this analysis are all those TIMSS countries that used more than 25 percent pedagogical suggestion units for their Population 2 science curriculum guides *and* that also had at least 20 blocks in the pool of their pedagogical suggestion units. Of these countries, several showed 100 percent consistency in using pedagogical suggestion blocks within pedagogical suggestion units. Only eight countries had an agreement of 50 percent or less: New Zealand, Singapore, Scotland, the Russian Federation, France, French-speaking Belgium, Hong Kong, and Iceland. They complemented the pedagogical suggestion blocks in the pedagogical suggestion units in varying ways: with example blocks in the case of the Russian Federation and Iceland, and a sizable proportion of assessment suggestions in Singapore and New Zealand.

The strategy of providing pedagogical support and facilitation obviously was not simple. Some countries, such as Japan and Israel, did not mix them with other unit types. Others, for example, Hungary, included a sizable proportion of examples.

Science Content Specified in Curriculum Guides

A fundamental question about the science curriculum guides sampled was to what extent specific science contents were included in those guides.

Inclusion of Science Content

Whether science content was specified varied for different block types. For each population, about one-half of the blocks in introductory and policy units had science codes. In the case of advanced physics, slightly less than one-half the blocks in introductory material had science content specified. However, for objective and content units, 95 percent or more of the blocks had science codes in all three populations. For pedagogy units, from about 80 to slightly less than 90 percent of the blocks had science content codes. The percentage was also high for other units.

Overall, the science curriculum guides specified science content with great frequency. This, unexpectedly, was less true in introductory and general policy units. Science content was almost always specified in objective and content units — as would be expected in documents for communicating science curricular goals. Pedagogical suggestions may vary from more general strategies and approaches to those that are specific to particular science contents. In the sampled guides, it is apparent that the focus was much more on particular, topic-specific pedagogical comments rather than more general comments — although some of the latter were clearly made. There was some variation among countries, but what was true overall was largely true for most countries individually. These were indeed and specifically *science* guides — and not general documents about curriculum.

Figure 4.3

Pedagogical Blocks within Pedagogy Units.

These are the percentages of the three types of pedagogy blocks within the pool of blocks from pedagogy units for a representative sample of countries and science guides for the upper grade of Population 2. The structure of pedagogy units were varied in many countries.

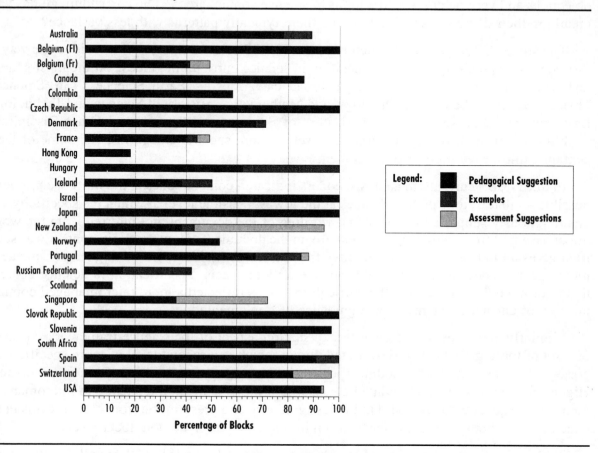

Specificity of Science Content

The sampled curriculum guides were largely devoted to communication about science topics. But how broadly or narrowly did they specify the science content and objectives? Did they provide broad, general goals and science content areas or much more specific instructional objectives and very precise specifications of science content areas? Did curriculum guides vary greatly in their degree of specificity or are there typically patterns with less variance?

The somewhat hierarchical structure of the TIMSS science framework provides one way of assessing how precisely content is specified in science curriculum guides. At the most general level, the science framework has eight broad categories — Earth Sciences; Life Sciences; Physical Sciences; Science, Technology, and Mathematics; History of Science and Technology; Environmental and Resource Issues; Nature of Science; and Science and Other Disciplines. Within each of these broad categories is a set of more specific categories, and some of these contain a third, even more specific content level.

In the TIMSS curriculum analysis, coders indicated content codes at the most specific level possible — that is, they provided codes at the lowest possible level at which the specificity was unambiguously supported by the actual wording of the document, not inferred from the wording or assumed from knowledge not actually in the document. How precisely content was specified gives an indication of how directive the curriculum guides were. A higher percentage of more specific codes assigned would indicate a more precise specification of science content; this in turn would indicate that the curriculum guides reflected a more micromanaged communication of curricular intentions and guidance of implementation.

Similarly, the student performance expectation aspect of the science framework consisted of a set of more global categories, nested within which is a second level of more specific categories. The instructions for coding specifically were the same as those for science content. Higher percentages of more detailed codes would indicate again a tendency to micromanagement. More general codes would imply less guidance and prescription from higher levels in the educational system — and greater freedom in implementation at more local levels.

Table 4.1 presents, for selected countries (see appendix I, table I4.1 for all countries), the percentage of blocks in all of each country's science curriculum guides that received at least one science content code at the most specific, middle, and most general levels of the science framework. The table also presents the percentage of blocks receiving performance expectation codes at the more specific and more general levels. As can be observed, countries varied considerably in the specificity of the statements concerning content and performance expectations in guides.

Some countries used the more general codes in describing the performances expected of students. Some countries — most notably Sweden and France for the upper grade of Population 1 — used mostly general specifications for the performances expected of students. The exact

Table 4.1
Specificity of Curriculum Guides.
These are the percentages for the guides of the upper grades of Populations 1 and 2 in a representative sample of countries showing whether content codes and performance expectation codes were assigned at very specific or more general levels of the TIMSS framework. Content, and to a somewhat lesser degree performance expectations, were generally given at specific levels.

	Population 1 - Upper Grade					Population 2 - Upper Grade				
	Content			Performance Expectations		Content			Performance Expectations	
Country	Most Specific	Mid-Level	Most General	Most Specific	Most General	Most Specific	Mid-Level	Most General	Most Specific	Most General
Austria	60	29	11	78	22	92	7	2	98	2
Dominican Republic	76	24	0	.	.	26	74	1	.	.
France	58	40	3	42	58	70	27	3	71	30
Greece	34	59	7	86	14	68	32	0	100	0
Israel	71	27	2	73	27	98	2	0	98	3
Korea	100	0	0	100	0	53	47	0	66	34
Mexico	87	13	0	100	0	97	3	0	100	0
Singapore[1]	53	18	29	56	44	44	38	18	75	25
Sweden	62	35	3	30	70	62	35	3	30	70
USA	85	13	2	92	8	84	14	2	97	3

1 The National Research Coordinator for Singapore notes that, while considering all unit types, there were a number of occasions in which the coders used mid-level or general codes; in all units devoted to the specification of content, coders used only the most specific codes.

significance of this is not clear. One possibility is that the curriculum guides themselves stated expectations for students at a more general level. Another possibility is that the science framework was inadequate for some countries to characterize their common expectations for students except in general terms; but the number of countries that did highly specify performance expectations suggests that this possibility is not the sole explanation.

Concluding Remarks

This chapter has offered an essential first step in the study of science curriculum guides. It makes clear that, based on an analysis of the sampled documents, there were considerable cross-national variations in the structure, specificity, and role of curriculum guides. In virtually all cases, however, they seem to have been carefully targeted documents, often of great specificity, by which curriculum designers communicated their intentions, especially to those shaping science curricula and providing instruction in actual classrooms.

NOTES

[1] 'Curriculum guide' is used here as a generic, standardized term for countries' official documents setting out policies and curricular intentions. They are called by many names in various countries - 'framework', 'syllabus', 'national curriculum,' and so on.

[2] The percentage of pages of a certain unit type rather than percentage of total units that were of that unit type had to be used for comparison here, even though the curriculum guide page sizes differ greatly among the documents, as does the amount of information per page (especially when the content of the pages are in different languages). Unit types for curriculum guides were not structured to be of comparable sizes as were unit types in textbooks. However, using percentage of pages for a given unit type expresses the proportion of a document devoted to a given unit type — that is, it compensates for different page sizes and linguistic densities since it compares pages to pages within the same document.

III. Reflections of Curriculum

Chapter 5

INTENDED FLOW OF SCIENCE CURRICULA

Science curricula "flow" over time. From the beginning of formal schooling to the end, science topics are introduced, receive attention for a time, and then stop receiving instructional time. This flow of science topics — those introduced, their duration, their departure — varies among countries and curricula, but the flow is planned and purposeful. It results from conceptions of how science should develop over schooling and of aims based on those conceptions.

A central issue in understanding curricular flow is the determination of whether science thinking develops independently of culture. One hypothesis assumes that there is a natural sequence to the development of scientific ideas that is linked to stages in human development; this sequence — and indeed, these stages — mostly holds true for all individuals and cultures. An opposing hypothesis considers science thinking — or any other aspect of thinking — to depend on culture. Perceptions, concepts, ways of knowing, and related aspects of thought are believed to differ significantly among cultures. The former position assumes a single appropriate sequence of science development that varies only slightly among countries, cultures, and languages. These variations represent either different understandings of when children arrive at readiness for particular topics or differences introduced accidentally by the decision making that creates science curricula and other plans of study. For the latter position, differences among countries in when science topics are introduced are far from accidental. The curriculum differences reflect important cultural differences as to how scientific thinking develops in a particular culture and how scientific knowledge is constructed and reconstructed socially.

This chapter looks at the differences and similarities across countries in curricular flow. The primary data source was judgments by panels of national experts and by ministry officials as informed by their country's curriculum guides. Their judgments concerned which topics in the Third International Mathematics and Science Study (TIMSS) science framework received attention in which years of science instruction. Appendix D details how they conveyed these judgments. As appropriate, experts also provided flow data for each science curriculum implemented in different regions, tracks, or other subnational segments of a country's educational system. They mapped in the broadest, least detailed sense the flow of educational experiences provided by schooling shaped by a nation's science curricula.

Given the wealth of data available for this analysis, a general picture of coverage for each science topic in the TIMSS framework is first created. Directing attention to the grade level at which a particular topic was introduced reduced but did not eliminate this diversity because each topic was introduced in differing grades in various curricula. This is an aggregation problem — how can one represent what was broadly common among diverse data yet not do it so broadly as to create a false impression? Here something like a median was used — in particular, the grade by which half of the TIMSS countries supplying curricular data had introduced a particular topic. Some countries will have introduced each such topic earlier, some later — but the grade level indicated represents when a topic was most typically introduced. By adding further data on the grade by which one-half of the countries had completed each topic's coverage, one has a broad portrait of how early and how long attention was devoted to each topic in the science framework.

Table 5.1 presents data on typical grades in which science framework topics were introduced and completed for the aggregate of TIMSS countries. In addition, data are given on the grade level at which the majority of all countries that covered a topic focused on the topic. These data are presented as if the flow of national curricula for each topic had been superimposed and a rough center for the introduction, conclusion, and focus had been determined. What emerges is a commonality — a "skeleton" of approximate similarity — for the place of each science topic in the composite flow of science curricula. Only further analyses will reveal whether this commonality is significant and substantive — or merely an accident due to the aggregation method.

The grade in which half the countries introduced topics varied from grade 2 to grade 11. Topics varied less in the grade by which they were concluded — from grade 10 to grade 12 in the aggregate. Obviously this created differences in the typical duration (number of years) to which curricular attention was devoted to a topic.

Introducing Science

Overall Picture

The TIMSS science framework offers a norm against which topic introduction can be compared. The lowest, most detailed levels of the TIMSS science framework orders topics within each category based on a priori judgment of approximate increasing difficulty.[1] The eight most general categories of the content aspect of the science framework represent overarching categories that organize the more detailed specifications that are possible with their subordinate codes. Organizing our conclusions according to these most general categories allows us to explore to some degree how various scientific disciplinary areas are intended for instruction.

- Most topics from the *earth and life* sciences tend to be introduced somewhere between grade 3 and grade 7 — the major exceptions being 'biochemical processes in cells' and topics in 'life spirals, genetic continuity, diversity,' some of which have not been

Table 5.1

Introducing, Focusing On, and Completing Science Topics.
There were important differences in the grade at which different science topics were introduced. There were also differences in the typical grade for completing topics and for focusing special attention on them.

Science Topics	Grade level at which half the countries had introduced the topic	Grade level at which the most countries focused on the topic	Grade level at which half the countries had completed the topic
Earth Sciences			
Earth Features			
Composition	5	10	11
Landforms	4	7	10
Bodies of Water	3	7	10
Atmosphere	5	7	11
Rocks, Soil	4	8	11
Ice Forms	7	8	12
Earth Processes			
Weather and Climate	3	7	10
Physical Cycles	4	7	10
Building and Breaking	5	7	11
Earth's History	6	12	12
Earth in the Universe			
Earth in the Solar System	4	7	10
Planets in the Solar System	5	9	11
Beyond the Solar System	7	9	11
Evolution of the Universe	10	12	12
Life Sciences			
Diversity, Organization, Structure of Living Things			
Plants, Fungi	2	7	11
Animals	2	7	11
Other Organisms	6	7	12
Organs, Tissues	3	8	12
Cells	6	10	12
Life Processes and Systems Enabling Life Functions			
Energy Handling	5	11	12
Sensing & Responding	7	12	12
Biochemical Processes in Cells	9	11	12
Life Spirals, Genetic Continuity, Diversity			
Life Cycles	3	12	12
Reproduction	3	9	12
Variation & Inheritance	8	12	12
Evolution, Speciation, Diversity	7	12	12
Biochemistry of Genetics	10	12	12

Table 5.1 (continued)
Introducing, Focusing On, and Completing Science Topics.

Science Topics	Grade level at which half the countries had introduced the topic	Grade level at which the most countries focused on the topic	Grade level at which half the countries had completed the topic
Interactions of Living Things			
Biomes & Ecosystems	5	7	12
Habitats & Niches	4	9	12
Interdependence of Life	4	8	12
Animal Behavior	3	7	11
Human Biology & Health			
Nutrition	2	8	12
Disease	3	8	12
Physical Sciences			
Matter			
Classification of Matter	5	10	11
Physical Properties	3	7	11
Chemical Properties	7	10	12
Structure of Matter			
Atoms, Ions, Molecules	7	10	12
Macromolecules, Crystals	9	11	12
Subatomic Particles	8	10	12
Energy and Physical Processes			
Energy Types, Sources, Conversions	5	11	12
Heat and Temperature	4	10	11
Wave Phenomena	9	11	12
Sound and Vibration	6	11	12
Light	5	12	12
Electricity	5	11	12
Magnetism	6	12	12
Physical Transformations			
Physical Changes	4	10	12
Explanations of Physical Changes	7	10	12
Kinetic Theory	10	10	12
Quantum Theory & Fundamental Particles	11	12	12
Chemical Transformations			
Chemical Changes	7	9	12
Explanations of Chemical Changes	9	9	12
Rate of Change & Equilibria	9	11	12
Energy & Chemical Change	9	11	12
Organic & Biochemical Changes	9	11	12
Nuclear Chemistry	10	12	12
Electrochemistry	9	10	12

Table 5.1 (continued)
Introducing, Focusing On, and Completing Science Topics.

Science Topics	Grade level at which half the countries had introduced the topic	Grade level at which the most countries focused on the topic	Grade level at which half the countries had completed the topic
Forces and Motion			
Types of Forces	7	10	12
Time, Space, & Motion	6	10	12
Dynamics of Motion	8	10	12
Relativity Theory	11	12	12
Fluid Behavior	11	9	12
Science, Technology, and Mathematics			
Nature or Conceptions of Technology	7	12	12
Interactions of Science, Mathematics, and Technology			
Influence of Mathematics & Technology in Science	10	12	12
Science Applications in Mathematics & Technology	7	12	12
Interactions of Science, Technology, & Society			
Influence of Science & Technology on Society	9	8	12
Influence of Society on Science, Technology	10	12	12
History of Science and Technology	8	11	12
Environmental and Resource Issues			
Pollution	4	9	12
Conservation of Land, Water, & Sea Resources	3	8	12
Conservation of Material & Energy Resources	5	8	12
World Population	7	12	12
Food Production, Storage	7	8	12
Effects of Natural Disasters	6	12	12
Nature of Science			
Nature of Scientific Knowledge	7	10	12
The Scientific Enterprise[1]			
Science and Other Disciplines			
Science & Mathematics	9	9	12
Science & Other Disciplines[1]			

1 So few countries intended to cover or focus on these topics that reporting any grade level would be misleading.

introduced in one-half of the TIMSS countries before grades 9 or 10. In earth science the sole exception is 'evolution of the universe.' Most topics introduced in grade 3 indicate an early primary curriculum mostly devoted to the study of taxonomy and weather. Once introduced, all of the life science topics remain in the curriculum until the end of upper secondary. This suggests that while topics are added to the curriculum with some regularity, mastery is not assumed until the conclusion of schooling. Perhaps it is the case that informal treatment at early grades is followed by formal treatment that concludes at the end of secondary.

- This pattern of the persistence of topics throughout the school years is apparently true for the *physical sciences* also. Although, most topics in this area are introduced later than was the case for life sciences (in secondary school — grades 7 to 12). They also are not instructionally completed in one-half of the countries until upper secondary, none before grade 11. This shows notable contrast from the pattern noted in mathematics curricula, where almost one-half of the topics introduced in primary school are completed by the end of lower secondary school. Only three topics are commonly introduced before grade 5: 'heat and temperature,' 'physical properties of matter,' and 'physical changes.'

- Only in the *earth sciences* are a substantial number of topics concluded by grade 10. Most topics are introduced before or at grade 6. A noteworthy exception is the cosmogony topic of 'evolution of the universe,' which one-half of the countries have introduced only by grade 10, perhaps as part of the treatment of astronomy and astrophysics in the physical sciences curricula.

The analysis of topic introduction considered which topics were introduced in the intended curriculum, within certain clusters of grades. Table 5.2 presents experts' judgments of intended introduction of science topics in a different arrangement. Here the topics are arranged by key groups of grades (that is, grades 1 to 3, grades 4 to 6, grades 7 and 8, and grades 9 to 12). While grades 1 to 3 and grades 4 to 6 clearly represented lower and upper elementary school it was less clear how — or whether — to separate grades 7 to 12, which generally represent secondary school. Grades 7 and 8 were particularly problematic. They are important transition grades and were grouped with different higher or lower grades in different countries. Because these grades cover the TIMSS targeted Population 2 of 13-year-old students, it was decided to designate them as a separate category in this analysis.

As table 5.2 shows, comparatively fewer topics are covered in the lower primary grades. Topics typically intended for introduction in grades 1 to 3 were mostly in life and earth sciences. They included a large proportion of topics involving classification, such as 'animals,' 'bodies of water,' 'plants, fungi,' and 'organs, tissues.'

More topics were intended for introduction in later elementary school science. In grades 4 to 6, content in the physical sciences was introduced, centering on topics of classification

Table 5.2
Science Topics Intended for Introduction in Various Stages.
Certain framework topics were typically introduced at various stages of schooling: grades 1 to 3, 4 to 6, 7 to 8, 9 to 12. This table presents a broad picture of the typical sequence of science topics introduction (representing the aggregate and not individual countries).

Grade Group	Topics That Half the Countries Intended for Introduction in Grade Group	
1 to 3	Bodies of water Weather & climate Plants, Fungi Animals Organs, Tissues Life Cycles of Organisms	Reproduction of Organisms Animal Behavior Nutrition Physical Properties of Matter Conservation of Land, Water, & Sea Resources
4 to 6	Composition Landforms Atmosphere Rocks, Soil Physical Cycles Building and Breaking Earth's History Earth in the Solar System Planets in the Solar System Other Organisms Cells Energy Handling Biomes & Ecosystems Habitats & Niches	Interdependence of Life Diseases Classification of Matter Energy types, Sources, Conversions Heat & Temperature Sound & Vibration Light Electricity Magnetism Physical Changes Time, Space, Motion Pollution Conservation of Material & Energy Resources Effects of Natural Disasters
7 to 8	Ice Forms Beyond the Solar System Sensing & Responding Variation & Inheritance in Organisms Evolution, Speciation, Diversity Chemical Properties of Matter Atoms, Ions, Molecules Subatomic Particles Explanations of Physical Changes Chemical Changes	Types of Forces Dynamics of Motion Nature or Conceptions of Technology Applications of Science in Mathematics, Technology History of Science & Technology World Population Food Production, Storage Nature of Scientific Knowledge

Table 5.2 (continued)
Science Topics Intended for Introduction in Various Stages.

Grade Group	Topics That Half the Countries Intended for Introduction in Grade Group	
9 to 12	Evolution of the Universe	Energy & Chemical Change
	Biochemical Processes in Cells	Nuclear Chemistry
	Biochemistry of Genetics	Electrochemistry
	Macromolecules, Crystals	Relativity Theory
	Wave Phenomena	Fluid Behavior
	Kinetic Theory	Influence of Mathematics, Technology in Science
	Quantum Theory, Fundamental Particles	
	Explanations of Chemical Changes	Influence of Science, Technology on Society
	Rate of Chemical Change & Equilibrium	Influence of Society on Science, Technology
	Organic & Biochemical Changes	Science & Mathematics

and transformation of matter and energy. Additional topics were introduced in the areas of environmental science, resource conservation, pollution, ecology, and ecosystems.

Grades 7 and 8 (key transition grades) saw introduction of mostly topics in physical science (especially chemistry) and technology. The topic of physical changes was expanded to include 'explanations of physical changes.' Additional environmental and resource areas, such as 'world population' and 'food production and storage,' were also introduced. For the first time, content on 'nature or conceptions of technology' and 'applications of science in mathematics and technology' was widely introduced across TIMSS countries.

Not surprisingly, grade 9 and beyond was characterized typically by the introduction of more advanced topics. Advanced topics in the life sciences included 'biochemical processes in cells' and 'biochemistry of genetics.' Mostly topics in the physical sciences are introduced, including 'relativity theory.' The earth science topic of 'evolution of the universe' is introduced also, most likely as part of astronomy in the physics curriculum.

Country Comparisons

Tables 5.1 and 5.2 characterize typical or "common denominator" curricular intentions across the science framework topics and the TIMSS countries, by presenting median grade levels for topic introductions. In actuality, some countries introduced any given topic earlier, others later. This section discusses these deviations from the median by individual countries.

Figure 5.1 presents, for all TIMSS countries, the number of topics introduced either 3 years earlier or later than the median. Note that this figure only addresses when topics were *introduced*, not duration or years of special focus.

Four countries stand out in this figure: Canada, Mexico, Portugal, and Slovenia. They differed by having more than 25 topics introduced three or more grades earlier than the median grade for that topic. This example of consistently introducing science topics earlier than the majority of TIMSS countries clearly reflects a very different distribution of opportunities to learn than that represented by the median. It is also likely to be reflected differently in educational attainments measured by the TIMSS achievement tests — to the extent that these tests reflect something like the profile of medians for science topic introduction — although this difference in scores would more clearly be seen for topics for which coverage in these four countries was a great deal earlier than typical for the test population grades. Such coverage differences would also be complicated by whether, after introduction, the curriculum in question focused more heavily on these topics than was typical. Such a consistent pattern is likely not due to accident. One might infer that these differences most likely reflect a particular curricular vision and are based on a deliberate set of aims underlying the science curriculum in these countries.

72 MANY VISIONS, MANY AIMS

Figure 5.1
Number of Topics Introduced Very Early or Very Late.
Some countries introduced many topics far earlier than was typical and some topics far later.

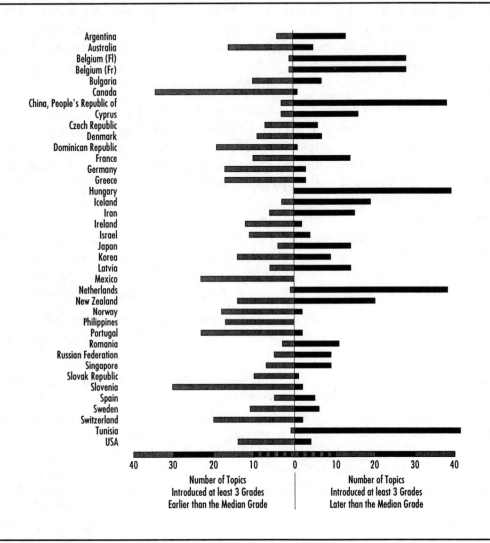

Figure 5.2

The Eight Topics Whose Introductory Grade Varied Most.

Many topics varied considerably among countries in terms of the grade in which they were introduced. These eight topics varied the most.

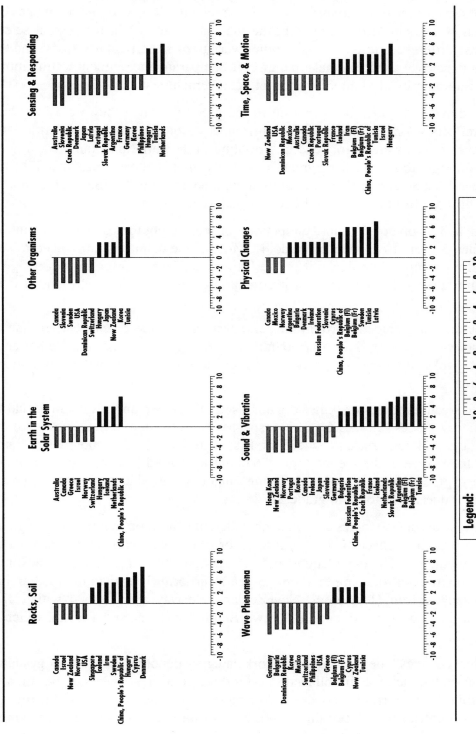

The patterns for the other countries were more mixed. Still, as figure 5.1 makes clear, many countries show marked differences from the median — and often for several science topics. Clearly these differences reflect cross-national variations in curricular visions, and their implementations in science curricula make the task of comparative attainment testing more difficult, and make caution critical to interpretation of achievement results.

Topics showed considerable variation in the grade in which they were introduced. Such variation reflects strong diversity in the curricular visions that helped determine the time in which science topics were introduced in schooling in the curricula examined. Figure 5.2 presents data for eight topics showing greatest diversity; that is, those topics that TIMSS countries introduced at least 3 years earlier or later than the median. Any TIMSS country not included in figure 5.2 differed from the median by less than 3 years for that topic.

'Rocks, soil' is one of the more elementary of these eight topics, and 13 countries differed by more than 3 years from median grade 4. Most of the countries introduced this topic later. Denmark, for example, introduced it in the 11th grade, 7 years after the median grade. Other countries introduced this material earlier, including Canada, which introduced it in the first grade.

Six countries introduced 'earth in the solar system' 3 or more years earlier than the median (which also was grade 4); only four countries — Hungary, Iceland, the Netherlands, and the People's Republic of China — introduced it 3 or more years later, perhaps in a more sophisticated fashion.

The life sciences topics of 'other organisms' and 'sensing and responding' had later median introductory grades - grades 6 and 7, respectively. Variation in introduction of these topics was quite large: 11 countries introduced the first of these topics 3 or more years from the median grade. In the case of the second topic, 16 countries varied 3 or more years from the median introductory grade, with the overwhelming majority introducing it earlier (Australia and Slovenia introduced this topic in grade 2).

Four physical science topics exhibited considerable variation in introductory grades. 'Wave phenomena' was introduced in 14 countries 3 or more years earlier or later than the median introductory grade 9. The topic 'sound and vibration,' exhibited perhaps the most notable variation: 22 countries introducing this topic 3 years earlier or later than the median introductory grade 6 — ranging from Hong Kong, New Zealand, Norway, and Portugal, which all introduced this material in grade 1, to Argentina, the two Belgian systems, and Tunisia, which introduced it in grade 12.

'Physical changes,' the science framework category denoting the study of gas laws, changes in states of matter, etc., was introduced in only three countries 3 years earlier than the median grade 4, while 13 countries introduced it 3 or more years later. In the case of 'time, space, and motion,' 18 countries introduced this material 3 or more years earlier or later than the median grade 6.

Duration of Topic Coverage

This sections looks at variations in how many grades various science topics were intended to be included in the curriculum. The *duration* data were derived by determining the number of grades during which curricular attention was given to each topic within each country. Duration must not be confused with instructional time. There is no reason to suppose that students in a country in which a particular topic is revisited and developed over a period of years are exposed to more instructional time in total than those where the topic is taught to mastery within a lesser number of years. The mean duration for each topic (across countries) and the mean duration for each country (across topics) was then calculated.

Unfortunately, there is no simple interpretation of average duration in an absolute sense. Some topics (for example, 'quantum theory') were widely covered in only a few grades. Other topics (for example, 'animal types') were widely covered over comparatively more grades. Any averaging over topics should be weighted by some index of typical duration for each topic. However, as seen earlier, the variations from such averages were considerable. The approach taken here was to find *unweighted* averages — each topic counting the same — and to use the data only for cross-national *comparisons*. Thus, an average duration of three grades would mean little in itself; it would, however, be more meaningful to compare two countries, one of which had an average duration three grades more than the other.

Overall Picture

After the average country durations were derived, the median duration was found among the pool of reporting countries. From this, the difference from that median was calculated for each country. As shown in figure 5.3, differences range from more than 3 years shorter than the median to more than 3 years longer. One group — Germany, the Philippines, and Singapore — were at or near the median duration. Others — most markedly Tunisia, the People's Republic of China, and Hungary — had an average duration less than the median; that is, on average, they devoted far fewer years than the median to covering a topic. At the other extreme, were countries like Canada, Slovenia, Switzerland, Portugal, and the Dominican Republic, which had comparatively high average durations. For some countries where the curriculum guides specify content only for ranges of grades and not for specific grade levels, such a policy would result in longer durations for some topics. Also, countries without national curriculum guides, but with multiple subsystems and their corresponding guides, could — when aggregated to the country level — produce longer durations for some topics. As a result, the country-level durations should not be interpreted as specific to individual students but rather to the country as a whole.

The average duration across all countries was computed for each framework topic. In this case, since each participating country is considered equally important, an unweighted average was not misleading. These topic averages represent the average number of grades devoted to each topic across the pool of reporting countries. Figure 5.4 shows that the average topic durations ranged from two grades ('quantum theory,' 'relativity theory,' and 'fluid behavior') to

Figure 5.3

Average Number of Grades Intended Across Topics by Countries.
Some countries had an average duration across topics that was far less than the median for all countries. Others had an average topic duration far greater than the median.

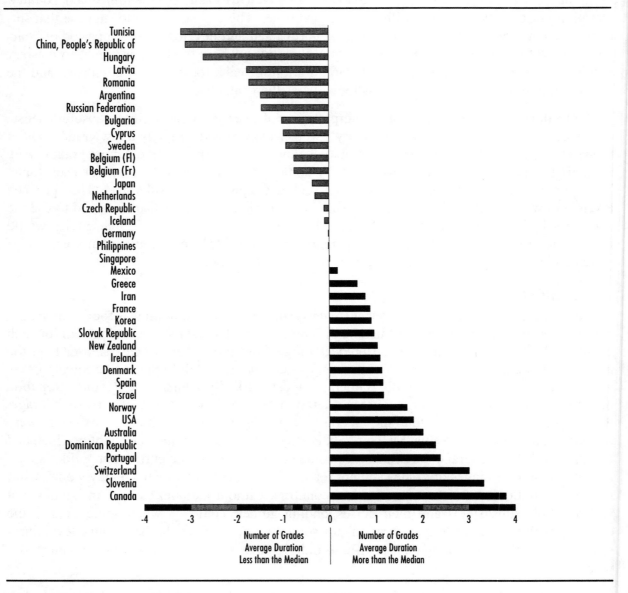

Figure 5.4

Average Number of Grades Intended for Each Topic.

More time was spent on some topics (longer average duration) than others, as these averages across all countries show. Other topics were covered for shorter times on average.

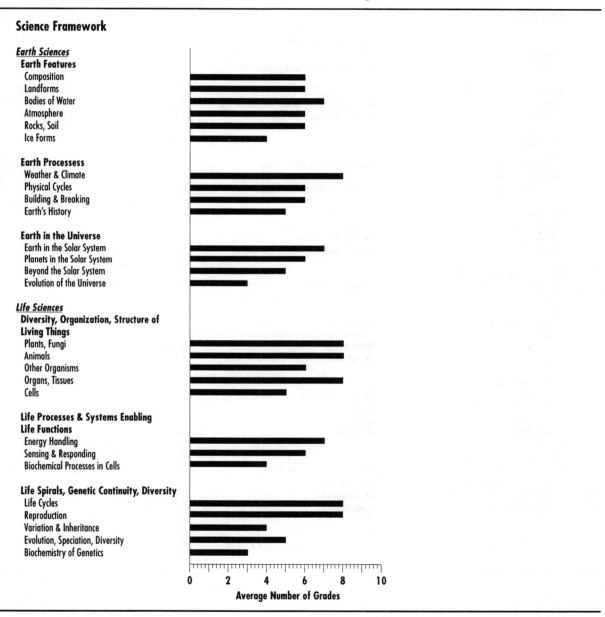

Science Framework

Figure 5.4 (continued)
Average Number of Grades Intended for Each Topic.

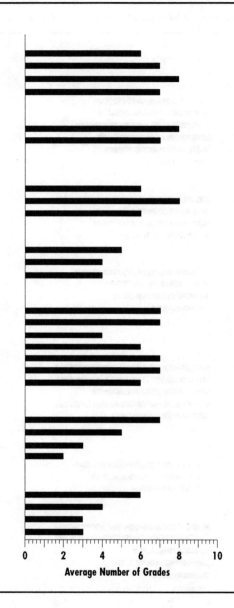

Organic & Biochemical Changes
Nuclear Chemistry
Electrochemistry

Forces & Motion
Types of Forces
Time, Space, & Motion
Dynamics of Motion
Relativity Theory
Fluid Behavior

Science, Technology, & Mathematics
Nature or Conceptions of Technology

Interactions of Science, Mathematics, & Technology
Influence of Mathematics, Technology in Science
Applications of Science in Mathematics, Technology

Interactions of Science, Technology & Society
Influence of Science, Technology on Society
Influence of Society on Science, Technology

History of Science & Technology

Environmental & Resource Issues
Related to Science
Pollution
Conservation of Land, Water, & Sea Resources
Conservation of Material & Energy Resources
World Population
Food Production, Storage
Effects of Natural Disasters

Nature of Science
Nature of Scientific Knowledge
The Scientific Enterprise

Science & Other Disciplines
Science & Mathematics
Science & Other Disciplines

0   2   4   6   8   10

Average Number of Grades

eight grades. Those topics with higher average durations were typically themes integrated into science curricula for comparatively longer spans or periodically revisited — even if there were particular grades when they were the focus of special attention.

Country Comparisons

Within the context of these two sets of average durations, there was considerable variation in the time devoted to any given topic by individual countries. As figure 5.5 shows, there was considerably more variation by individual topic and country than there was in either set of averages.

Some topics seem to have been commonly integrated into the science instruction of several grades or revisited periodically. For example, 'animals' and instruction related to 'magnetism' were involved in an average of eight and six grades, respectively. Yet there are noteworthy exceptions — the People's Republic of China in the case of 'animals' and the Dominican Republic, Israel, New Zealand, and Slovenia in the case of 'magnetism.'

Other topics, exhibited considerably less cross-national commonality. The study of 'effects of natural disasters' had an average duration of four grades. However, examining the duration of intended work on this topic in individual countries showed marked variation — from as few as one grade to as many as eight. For this topic and others, there appears to have been considerably less consensus about the duration of curricular attention devoted to them — and, by inference, about the approaches taken to them and the educational opportunities provided related to them. One approach for identifying these differences would be to see how intended curricular attention for a given science framework topic differed among the countries. Here, a more complex approach was used that examined *double differences*. Statistical techniques were used to see how the duration for a given topic in a given country differed *at the same time* from that typical for the topic (across countries) and for the country (across topics). This yielded a single number indicating how differently that topic was treated by that country from how a typical country treated a typical topic.

Figure 5.5 presents the data on these double differences for countries and topics that had many combined differences that were four or more grades greater or fewer than typical. The row of numbers indicate each *country's* typical topic duration and the column of numbers indicate each *topics* typical duration. Grey-shaded icons indicate fewer years than typically intended; black icons indicate more years than typically intended. As shown, there was considerable diversity revealed by this analysis; display of the data for all countries (see appendix I, figure I5.5) would make the overall diversity even more striking.

The 14 topics in figure 5.5 showed strong differences in many countries, some with considerably fewer intended for each topic. Countries also differed in how often large "double differences" emerge. In New Zealand, such differences emerged for seven topics that differed by 3 years or more. In Canada and Korea, only two differences of similar magnitude were present.

Figure 5.5

Specific Country/Topic Durations Compared to What Was Typical.

Statistical methods were used to produce numbers indicating how many fewer (gray-shaded icons) or more (black-shaded icons) grades the duration for a specific country and topic was from the duration for a typical country and topic — that is, a double difference. Considerable diversity is shown.

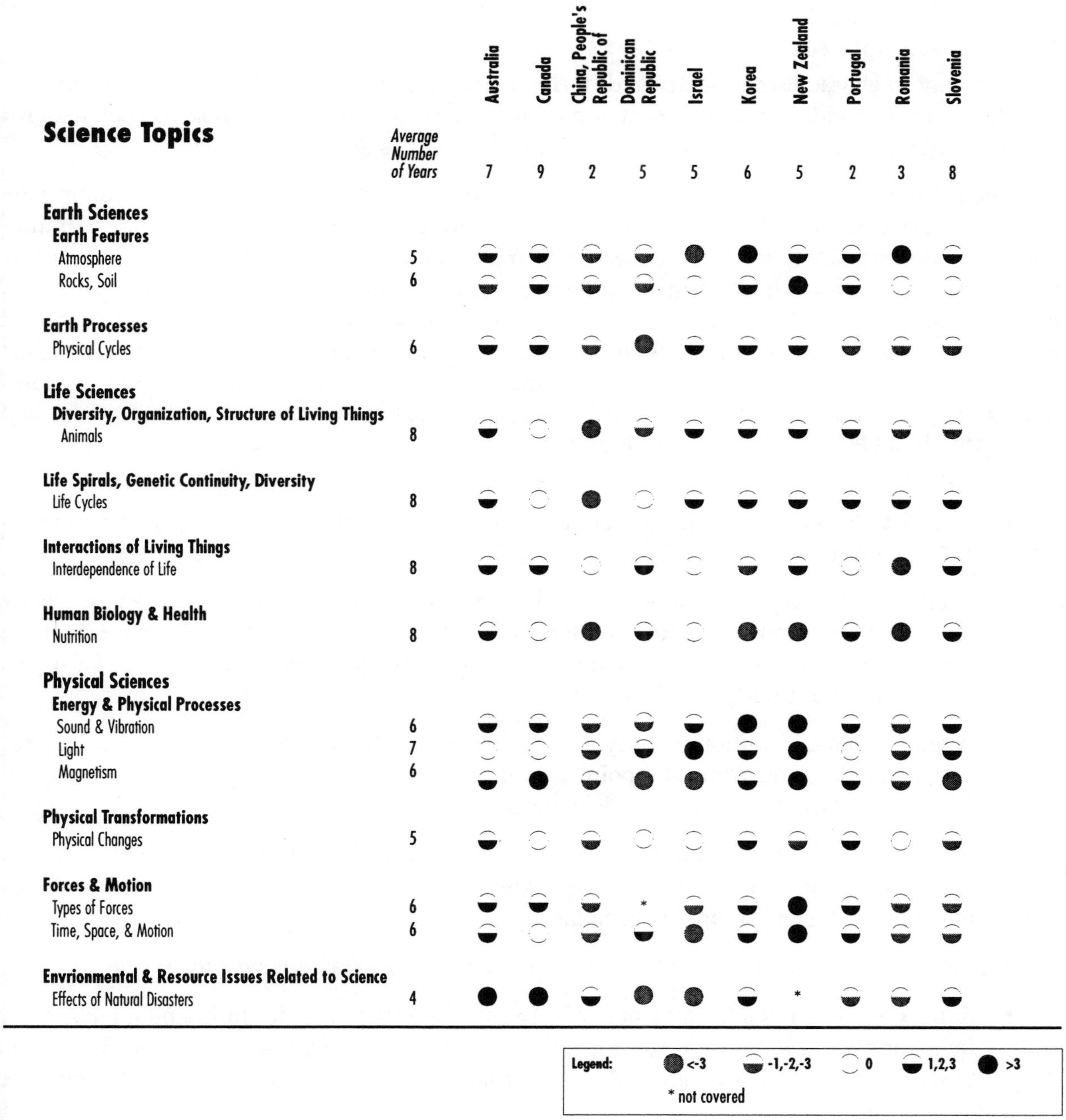

Science Topics	Average Number of Years	Australia (7)	Canada (9)	China, People's Republic of (2)	Dominican Republic (5)	Israel (5)	Korea (6)	New Zealand (5)	Portugal (2)	Romania (3)	Slovenia (8)
Earth Sciences											
Earth Features											
Atmosphere	5										
Rocks, Soil	6										
Earth Processes											
Physical Cycles	6										
Life Sciences											
Diversity, Organization, Structure of Living Things											
Animals	8										
Life Spirals, Genetic Continuity, Diversity											
Life Cycles	8										
Interactions of Living Things											
Interdependence of Life	8										
Human Biology & Health											
Nutrition	8										
Physical Sciences											
Energy & Physical Processes											
Sound & Vibration	6										
Light	7										
Magnetism	6										
Physical Transformations											
Physical Changes	5										
Forces & Motion											
Types of Forces	6										
Time, Space, & Motion	6										
Environmental & Resource Issues Related to Science											
Effects of Natural Disasters	4										

Legend: ● <-3 ◓ -1,-2,-3 ○ 0 ◒ 1,2,3 ● >3

* not covered

The flow of science curricula in the TIMSS countries differed not only in when individual topics were introduced into their science curricula, but also in terms of how long topics continued to receive curricular attention (duration). These differences in placement and topic coverage must have consequences for educational attainments and would seem to be important factors to consider in accounting for such differences.

Covering Multiple Topics

One way to characterize countries' differing curricular patterns is to examine how curricular attention was divided among science topics at each grade. Figure 5.6 presents such data for various TIMSS countries. Certain cross-national differences are readily apparent.

One broad distinction is between more focused and more diverse curricula. Examining the average number of topics across the grades, the picture that emerges is one of broad, simultaneous — within a single grade — coverage of many topics or of attention restricted to a few topics. The data make clear that most countries presented an additive or "fanning" pattern of curricular intentions. In each of these countries, a comparatively small number of topics received attention in early grades, with new topics being introduced steadily each year. This pattern contrasts markedly with the pattern found for mathematics, which showed the predominance of a "curvilinear" pattern: one in which few topics are introduced in early grades and progressively added to until some midpoint, at which progressively fewer topics were intended. This resulted in a "bell-shaped" or "normal-curve" pattern.

There appear to be three principal variations on the predominating additive pattern in curricula for the sciences.

- "Purely" additive curricula — countries like Australia in which there was a steady increase in the number of topics intended in each successive grade. In grade 1, these curricula showed the smallest number of topics intended; the largest number of topics appeared in the final grade of upper secondary.

- A higher rate of addition in early grades reaching a "saturation" point early in the secondary grades, and from this point onward a fairly consistent high number of topics. Cyprus and Greece exemplify this pattern.

- A slight reversal of the additive pattern in the final grades of upper secondary, such as in Norway and Sweden, where there was a slight reduction in the number of topics intended at the end of upper secondary.

In addition to the prevailing additive pattern, there were a few noteworthy exceptions:

- Some countries intended a comparatively small number of topics throughout the grades, indicating more focus on fewer topics. This pattern resulted in a "leaner" profile, and is best exemplified by Argentina. Another group of countries, which included the People's

Figure 5.6

Number of Topics to Be Covered for Each Grade in Each Country.

The number of topics to be covered in various grades differs considerably across countries.

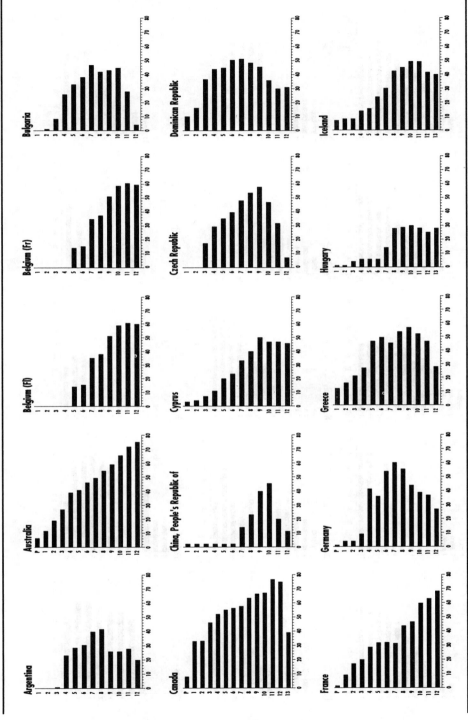

Note: Data for the secondary level within countries pertain to the general track only.

Figure 5.6 (continued)
Number of Topics to Be Covered for Each Grade in Each Country.

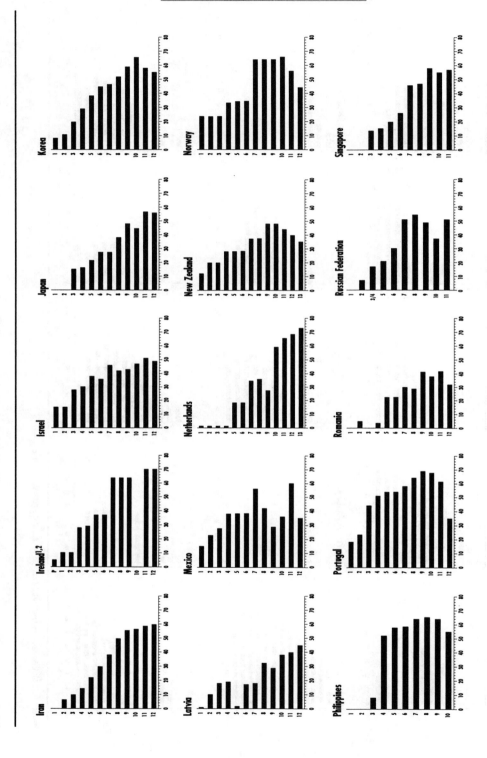

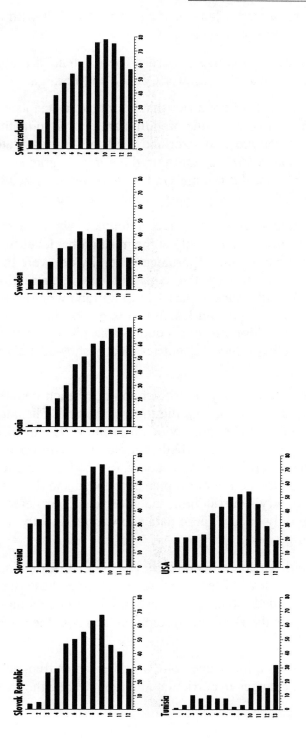

Note: Data for the secondary level within countries pertain to the general track only.

1 Ireland's two pre-primary grades (ages 5 and 6) were combined into one grade.

2 Grade 10 is an optional transitional year in Ireland; there is no national curriculum prescribed for this year.

Republic of China and Hungary, also presented a lean or focused profile, but with an increase in the number of topics intended in upper secondary.

- Another group presented a more consistently diverse pattern. These intended a high number of topics throughout the grades, as was the case for Canada and Portugal.

- Finally, another noteworthy exception was that of countries that presented in science the same curvilinear pattern of curriculum discovered as the predominant pattern in mathematics. Thus, the Czech and Slovak Republics, Switzerland, and the United States exemplify countries that started out with an *additive* pattern until the middle grades and then presented a *subtractive* pattern such that the middle grades were characterized by intentions to cover more topics than in either lower primary or upper secondary.

The data presented in figure 5.6 are descriptive — and suggestive. Inferences about the distribution of instructional attention among topics can be stated only at the most global level from these data — not all science framework topics are of equal importance generally nor were they within any given reporting country. Not all topics require the same amount of instruction. However, at a global level, a country that deals with many topics within a grade would seem likely to differ in the organization of its educational opportunities, in its textbooks and curricular materials, and in its curricular intentions and visions from a country that clearly restricts attention — even at this global level — to only a few topics. Countries clearly showed a diversity of curricular visions in this way.

Figure 5.6 portrays how topics accumulated over the years of schooling in each country. After they entered the science curriculum of a particular country, did they ever exist? The additive pattern seen for many countries in figure 5.6 provides no obvious indication. The data for how many topics received attention do not directly portray whether topics were dropped at some point. They also do not indicate how many topics were introduced for the first time at a given grade. Figure 5.7 combines information on the number of topics introduced and the number of topics dropped at each grade within a country's curriculum. Each topic is considered to be introduced or dropped from the curriculum only once. These data show more directly how diversity was created and how it changed over grades.

Some countries regularly added topics and dropped comparatively few — that is, their science curricula grew incrementally over the grades. Clearly included among these were Australia, France, and Spain. Others regularly added topics in the early grades and then mostly dropped topics. Most notable among these were the Dominican Republic and both the Czech and Slovak Republics.

Of particular interest are those countries that both introduced *and* dropped topics frequently — that is, countries with comparatively constant changes in curricular attention. While the over-

Figure 5.7
Number of Topics Introduced and Dropped.

Countries differed in terms of how many topics were introduced and how many were dropped at each grade. These data show how the number of topics covered in each grade were accumulated.

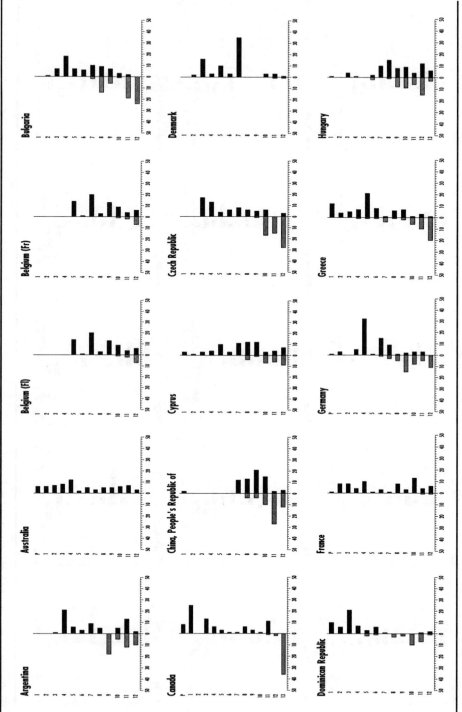

Note: Data for the secondary level within countries pertain to the general track only.

Figure 5.7 (cont'd)
Number of Topics Introduced and Dropped.

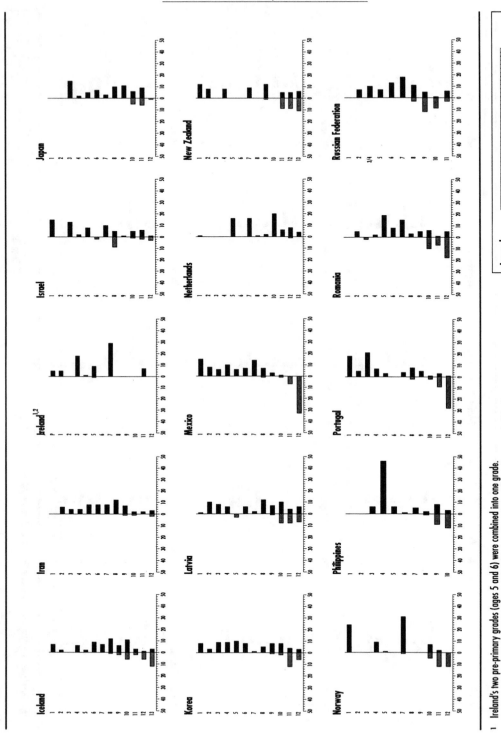

1 Ireland's two pre-primary grades (ages 5 and 6) were combined into one grade.

2 Grade 10 is an optional transitional year in Ireland; there is no national curriculum prescribed for this year.

Note: Data for the secondary level within countries pertain to the general track only.

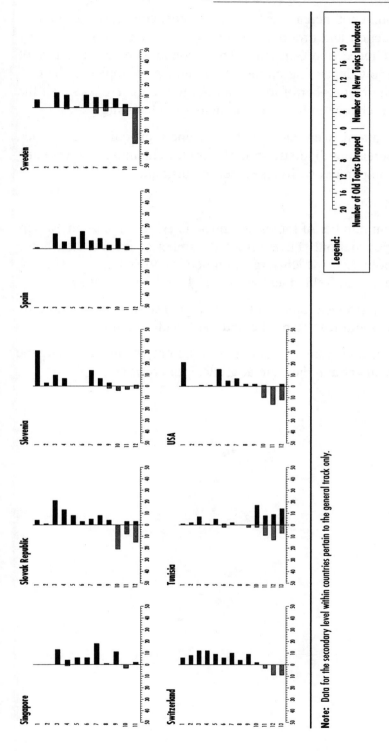

Note: Data for the secondary level within countries pertain to the general track only.

Legend:

| Number of Old Topics Dropped | Number of New Topics Introduced |

all multiplicity of topics attended to might not increase as much as in other countries, the changes of topic indicated by frequent introduction and dropping of topics were clearly comparatively great. This group includes Hungary, Greece, and Germany. These countries reveal a pattern of changing focus — either broad or narrow depending on the actual number of topics in the curricular stream — of devoting attention to a number of topics but not to the *same* topics over the years. Countries with a high degree of change in focus are particularly interesting.

Thus, overall, three predominant patterns were identified for science curricula — continued progressively more diverse focus, increasing diversity in early grades changing to increasing focus in later grades, and changing comparatively diverse science curricula.

Some other patterns worth noting:

- Countries in which a significant number of topics are introduced only in certain "watershed" grades. That is, countries that only have a limited number of grades, distributed throughout the years of schooling, in which large numbers of topics are introduced. Such is the case of grades 3 and 7 in Ireland and grades 1, 4, and 7 in Norway.

- Countries that seem to mostly introduce topics in the early grades, dropping them only in later grades, such as in the Dominican Republic and the Czech Republic.

- The People's Republic of China and Hungary stand out as the only countries where the introduction of new topics occurs at a higher rate at grade 7 and beyond.

NOTE

[1] This ranking applies only at the most detailed level; the eight highest level categories (for examples, earth science, life sciences, physical sciences, etc.) are not listed in any order.

Chapter 6

MILESTONES IN SCIENCE CURRICULA

So far, this analysis has focused on broad patterns in the flow of science curricula — patterns of intended topics across the span of grades. Such proposed coverage was not evenly distributed and varied considerably from grade to grade. The data from the Third International Mathematics and Science Study (TIMSS) allowed examination of the focus of intended coverage at each grade. Content covered in curriculum guides and textbooks for selected grades can also be examined in some detail. This chapter examines the distribution of focus topics across grades and, in some detail, in the upper grades of the TIMSS target populations. This includes examination of the match between intended coverage as expressed in curriculum guides and textbooks. Finally, there is an attempt to characterize the broadest level of content commonality in those target grades by examining specific commonly intended topics.

Patterns of Topic Focus

Much of the data used thus far were gathered by surveying science curriculum experts in each country, framework topic by topic. These experts used national curriculum guides to provide data on the grades in which various science topics were introduced and concluded and, thus, information on the coverage duration for the various topics. These experts were also asked to provide information on the grades within which comparatively greater focus was given to particular science topics. These data, although drawn from informed expert judgment rather than document analysis, provide an initial, direct look at whether and when particular science topics received special attention and focus in the curricular flow of each country (see table 5.1 in chapter 5).

Broad commonalities emerge for the grades in which particular science topics were indicated for intended special focus. Some topics of comparatively less significance had no focus indicated by a majority of countries, but most did. These commonalities in topic focus make up a "common denominator" focused curriculum, a curriculum made up of the sequence of those topics that were indicated as receiving particular focus in the various grades.

Several topics of focus appeared in grades 7 and 8. Among these were most earth science topics from the science frameworks and also the most elementary life sciences topics from the topic areas of 'diversity, organization, structure of living things,' 'life spirals,' and some elements of environmental science, ecology, and animal and plant behavior. The only physical sciences focused on in these grades by most countries are 'heat and temperature' and 'physical changes.'

In grades 9 and 10, physical science became a major focus. Some topics in environmental and resource issues related to science were also focused on in these grades.

A final set of focus topics were uncovered for grades 11 and 12. These included work on the majority of topics in 'chemical transformations' and a focus on the majority of 'energy and physical processes' topics.

Similarities of curricular focus were in a context of persistent — and at times large — diversity. Based on the common denominator focus pattern of table 5.1 for selected, representative countries and science topics for which the diversities were more marked, figure 6.1 presents most of the countries and topics that showed differences of grade levels from the majority focus grade.

Certain topics showed differences from the majority focus grade more often. 'Rocks, soil' was a focus for four or more grades earlier in at least two countries — Canada and Israel — indicating that there were countries that shared the characteristic of early focus on such material. Material on biological variation and inheritance was also focused on comparatively earlier in Denmark, Mexico, and the Slovak Republic. Denmark and Israel also focused on 'evolution, speciation, and diversity' topics very early.

In addition to science topics that showed some large differences, some countries showed a number of differences from the common denominator profile of science topic focus grades. For example, for topics in the earth and life sciences as well as in the category of 'energy and physical processes,' Israeli experts indicated a much earlier focus in contrast to the more common, majority focus grades.

Topics in the physical sciences showed interesting variation in patterns of focus. Grade 11 was the point at which most countries focused on the topic of 'electricity.' Some countries, however, focus on this topic in lower elementary school from 8 to 10 years prior to grade 11 (i.e., Israel and Japan). Some countries waited until the final grade of upper secondary school to focus on this topic, as was the case for Flemish-speaking Belgium.

The data from table 5.1 clearly indicated that there were some broad commonalities in the grades in which many science framework topics received special attention. However, figure 6.1 shows that there were also marked diversities (see appendix I, figure I6.1 for all countries). Certain topics — for example, 'light' — received special attention in several countries at characteristically early times. Certain countries — for example, Israel — show a pattern of frequently having focused on at least some contents earlier than did the majority. The pattern of topic focus for individual countries and topics was one of considerable cross-national diversity.

Figure 6.1

Focus Differences from the Majority for Selected Countries.

For most topics there was one grade by which a majority of countries had focused on the topic. Individual countries had foci that differed from the majority for at least some topics. A few countries were selected to illustrate the differences in which focus differed from the majority. Negative numbers (gray-shaded icons) indicate focus earlier than was typical. Some differences were large enough to indicate quite different approaches to the topics involved.

Science Topics	Grade level at which the majority of countries focused on a topic	Belgium (Fl)	Canada	China, People's Republic of	Denmark	Hungary	Ireland	Israel	Japan	Korea	Mexico	Portugal	Russian Federation	Slovak Republic
Earth Sciences														
Earth Features														
Rocks, Soil	9													
Earth Processes														
Weather & Climate	7													
Earth in the Universe														
Earth in the Solar System	7													
Life Sciences														
Diversity, Organization, Structure of Living Things														
Organs, Tissues	8													
Life Spirals, Genetic Continuity, Diversity														
Variation & Inheritance	12					*								
Evolution, Speciation, Diversity	12					**								
Physical Sciences														
Matter														
Classification of Matter	10				*									
Structure of Matter														
Atoms, Ions, Molecules	10													
Energy & Physical Processes														
Light	12													
Electricity	11													
Chemical Transformations														
Chemical Changes	9				*									
Environmental & Resource Issues														
Pollution	9						*							

Note: In those instances where there is more than one focus year, the earlier one was used in creating this table.

Legend: ● <-4   ◗ -1 to -4   ○ 0   ◓ 1 to 4   ● >4

* not focused

Commonly Intended Topics Across Grades

The portraits drawn in chapter 5 and earlier in this chapter sought to capture the flow of a common denominator curriculum and to illuminate the background diversity in the details of particular topics and countries that lay behind commonly intended coverage. The broadest findings were clear — strong common features amid considerable diversity. The data were often portrayed to clarify differences among individual countries and topics. While this presentation was necessary, it has made it hard thus far to assess the degree of agreement among the countries — how common the different common denominator factors were in content coverage and content focus. Data are presented in this section in an effort to remedy this problem.

What proportion of the TIMSS countries covered a given science framework topic in a given grade? To answer this question in detail would require a two-dimensional grid of topics by grades and a percentage in each cell of the grid — each combination of topic and grade — indicating what proportion of the countries cover that topic in that grade. Such data require simplification if any sense of the underlying patterns is to emerge. An important simplification here — a reduction in the complexity portrayed — is not to portray every percentage numerically. Thus, we empirically identified a limited set of percentage levels that captured important differences among topics and countries.

Through the identifying process, we found that 70 percent and 10 percent were important milestones in commonality, partitioning countries and topics in a way that revealed important differences in content coverage. Four categories were used:

- More than 70 percent of the countries covered the given topic at the given grade level;

- From 10 to 70 percent of the countries covered it;

- Some countries, but fewer than 10 percent covered it; and

- No countries covered the given topic at the given grade level.

While the choice of 10 percent and 70 percent was arbitrary, it was based on careful analysis to determine which cutoff percentages would best create a small, illuminating set of coverage levels.

Figure 6.2 presents these data on content coverage for the full range of science framework topics and for the full range of grades. Once we identified the four categories described above, a set of icons were created to portray these categories — essentially high, moderate, low, and no agreement across the pool of countries. This is an indication of common curricular intention, not of focus. Here, inclusion was determined simply by whether a country indicated that a topic received some attention during the grade in question. These patterns present the broad outlines of which topics were commonly intended cross-nationally.

The number of filled icons — indicating at least 70 percent coverage — is relatively small. Only a few regions of filled icons emerge to indicate more inclusive content categories and spans of grades with broad commonality. For example, such regions include work with 'plants, fungi' and 'animals' from grade 3 to grade 8; 'pollution' and 'resource conservation' in grades 7 to 11; 'life cycles of organisms' and 'reproduction of organisms' in grades 5 to 11; and various aspects of physical and chemical properties of matter from grades 7 and 8 up to grades 10 and 11.

The patterns that emerge are quite striking. In the earth sciences, the lowest grade at which 70 percent or more countries covered a given topic was for 'bodies of water,' 'weather and climate,' 'physical cycles,' and 'earth in the solar system' at grades 4 and 5. Most of the topics in this area are covered by the majority of countries in upper elementary and lower secondary school.

Life science is also generally emphasized in upper elementary and lower secondary school. Plant and animal taxonomy dominate elementary and lower secondary curricula (grades 3 through 8), with topics such as 'organs, tissues' being largely covered at the same time. Ecological topics such as 'biomes, ecosystems,' 'habitats, niches,' and 'interdependence of life' are covered by most countries in lower secondary school. However, at least 70 percent of the countries covered the latter topic throughout most of lower and upper secondary school.

Physical sciences were covered later: 70 percent or more of the TIMSS countries covered most of this framework area in grades 7 and 8, starting with 'heat and temperature' in grade 6. By grade 8, 70 of TIMSS countries were covering nine physical science topics from the general categories of 'matter,' 'structure of matter,' energy and physical processes, and 'physical transformations.'

Environmental and resource issues related to science were commonly covered in grades 7 through 11.

A similar — but more finely detailed — portrait can be drawn by providing an analogous display for indicated *focus* on, rather than coverage of a particular topic in a particular grade. Figure 6.3 presents focus data for the full range of science framework topics and for the full range of grades. In this analysis, inclusion in computing a percentage was determined by whether a country indicated that a topic received a focus during the grade in question. The cutoffs (10 and 70 percent), four common intention levels, and icons are the same as those used in figure 6.2.

There are no filled icons in the entire figure. Many of the filled icons from figure 6.2 are now shaded to indicate moderate agreement at the 10 to 70 percent level — as are many of the icons that were already at this moderate agreement level in the consideration of coverage only. Achieving a high degree of common intention at the focus level would indicate more serious common commitments of instructional attention. Topics that did achieve a high level of commonly attended focus would be, more precisely, the cross-national commonalities for which comparable attainment measures would be most appropriate at the grades indicated. It is perhaps in this sense that figure 6.3 is most cautionary — and disheartening. No matter what set of grades was selected, there would be no topics within it for which more than 70 percent of the countries indicated an intention to focus instructional attention.

Figure 6.2
Proportions of Countries Covering Each Topic at Each Grade.

The actual percentages of countries covering each topic at each grade were used to identify a coverage level indicated by one of four symbols. The display makes it easier to identify topics and grades for which there were strong similarities in content coverage.

Science Topics	P	1	2	3	4	5	6	7	8	9	10	11	12
Earth Sciences													
Earth Features													
Composition	○	○	◑	◑	◑	◑	●	●	◑	●	◑	◑	◑
Landforms	○	◑	◑	◑	◑	◑	●	●	●	◑	◑	◑	◑
Bodies of Water	○	◑	◑	◑	◑	●	●	●	●	●	◑	◑	◑
Atmosphere	○	○	◑	◑	◑	◑	◑	◑	◑	◑	◑	◑	◑
Rocks, Soil	○	◑	◑	◑	◑	◑	◑	◑	◑	●	◑	◑	◑
Ice Forms	○	○	○	◑	◑	◑	◑	◑	◑	◑	◑	◑	◑
Earth Processess													
Weather & Climate	○	◑	◑	◑	●	●	●	●	●	◑	◑	◑	◑
Physical Cycles	○	◑	◑	◑	●	●	●	●	●	◑	◑	◑	◑
Building & Breaking	○	○	○	◑	◑	◑	◑	◑	●	●	◑	◑	◑
Earth's History	○	○	○	○	◑	◑	◑	◑	◑	◑	◑	◑	◑
Earth in the Universe													
Earth in the Solar System	○	◑	◑	◑	◑	◑	●	●	●	◑	◑	◑	◑
Planets in the Solar System	○	○	○	◑	◑	◑	◑	●	◑	◑	◑	◑	◑
Beyond the Solar System	○	○	○	○	◑	◑	◑	◑	◑	◑	◑	◑	◑
Evolution of the Universe	○	○	○	○	◑	◑	◑	◑	◑	◑	◑	◑	◑
Life Sciences													
Diversity, Organization, Structure of Living Things													
Plants, Fungi	○	◑	◑	●	●	●	●	●	●	◑	◑	◑	◑
Animals	○	◑	◑	●	●	●	●	●	●	◑	◑	◑	◑
Other Organisms	○	◑	◑	◑	◑	◑	◑	●	●	◑	◑	◑	◑
Organs, Tissues	○	◑	◑	◑	◑	●	●	●	●	◑	◑	◑	◑
Cells	○	○	○	○	◑	◑	◑	●	●	●	●	●	◑
Life Processes & Systems Enabling Life Functions													
Energy Handling	○	○	◑	◑	◑	◑	◑	●	●	●	●	●	◑
Sensing & Responding	○	○	○	◑	◑	◑	◑	◑	●	●	◑	◑	◑
Biochemical Processes in Cells	○	○	○	○	◑	◑	◑	◑	◑	◑	◑	●	◑
Life Spirals, Genetic Continuity, Diversity													
Life Cycles	○	◑	◑	◑	◑	●	●	●	●	●	●	●	◑
Reproduction	○	◑	◑	◑	◑	●	●	●	●	●	●	●	◑
Variation & Inheritance	○	○	○	○	○	◑	◑	◑	◑	●	●	●	◑
Evolution, Speciation, Diversity	○	○	○	◑	◑	◑	○	◑	◑	●	●	●	◑
Biochemistry of Genetics	○	○	○	○	◑	○	○	◑	◑	◑	◑	◑	◑

Science Topics

	P	1	2	3	4	5	6	7	8	9	10	11	12
Interactions of Living Things													
Biomes & Ecosystems	○	○	◐	◐	◐	◐	◐	●	●	◐	●	●	◐
Habitats & Niches	○	◐	◐	◐	◐	◐	◐	●	●	◐	◐	●	◐
Interdependence of Life	○	◐	◐	◐	◐	●	●	●	●	●	●	●	◐
Animal Behavior	○	◐	◐	◐	◐	◐	◐	◐	◐	◐	◐	◐	◐
Human Biology & Health													
Nutrition	○	◐	◐	◐	◐	◐	●	●	●	◐	◐	◐	
Disease	○	◐	◐	◐	◐	◐	◐	●	●	●	●	●	
Physical Sciences													
Matter													
Classification of Matter	○	◐	◐	◐	◐	◐	◐	●	●	●	●	◐	◐
Physical Properties	○	◐	◐	◐	◐	◐	◐	●	●	●	●	◐	◐
Chemical Properties	○	○	○	○	◐	◐	◐	◐	●	●	●	●	◐
Structure of Matter													
Atoms, Ions, Molecules	○	○	○	○	○	◐	◐	◐	●	●	●	●	◐
Macromolecules, Crystals	○	○	○	○	○	○	◐	◐	◐	◐	●	●	◐
Sub-atomic Particles	○	○	○	○	○	○	◐	◐	◐	●	●	●	◐
Energy & Physical Processes													
Energy Types, Sources, Conversions	○	◐	◐	◐	◐	◐	●	●	●	●	●	●	◐
Heat & Temperature	○	◐	◐	◐	◐	◐	●	●	●	●	●	●	◐
Wave Phenomena	○	○	○	◐	◐	◐	◐	◐	◐	◐	●	●	◐
Sound & Vibration	○	○	◐	◐	◐	◐	◐	◐	◐	◐	●	●	◐
Light	○	◐	◐	◐	◐	◐	◐	◐	●	●	●	●	◐
Electricity	○	○	◐	◐	◐	◐	◐	●	●	●	●	●	◐
Magnetism	○	◐	◐	◐	◐	◐	◐	◐	◐	◐	●	◐	◐
Physical Transformations													
Physical Changes	○	○	◐	◐	◐	◐	◐	●	●	●	●	●	◐
Explanations of Physical Changes	○	○	○	○	◐	◐	◐	◐	◐	◐	●	●	◐
Kinetic Theory	○	○	○	○	○	○	○	◐	◐	◐	◐	◐	◐
Quantum Theory & Fundamental Particles	○	○	○	○	○	○	○	○	○	◐	◐	◐	◐
Chemical Transformations													
Chemical Changes	○	○	○	○	◐	◐	◐	◐	●	●	◐	●	◐
Explanations of Chemical Changes	○	○	○	○	○	◐	◐	◐	◐	◐	◐	●	◐
Rate of Chemical Change & Equilibrium	○	○	○	○	○	○	○	◐	◐	◐	◐	◐	◐
Energy & Chemical Change	○	○	○	○	○	○	○	◐	◐	◐	◐	◐	◐

Legend:

● at least 70% of the countries covered this topic at this grade level

◐ 10 to 70% of the countries covered this topic at this grade level

◔ at most 10% of the countries covered this topic at this grade level

○ no country covered this topic at this grade level

Figure 6.2 (continued)
Proportions of Countries Covering Each Topic at Each Grade.

Science Topics	P	1	2	3	4	5	6	7	8	9	10	11	12
Organic & Biochemical Changes	○	○	○	○	○	○	○	◐	◐	◐	◐	●	◐
Nuclear Chemistry	○	○	○	○	○	○	○	◐	◐	◐	◐	◐	◐
Electrochemistry	○	○	○	○	○	○	○	◐	◐	◐	◐	●	◐
Forces & Motion													
Types of Forces	○	◔	◔	◐	◐	◐	◐	◐	◐	●	●	●	◐
Time, Space, & Motion	○	◔	◐	◐	◐	◐	◐	◐	●	●	●	●	◐
Dynamics of Motion	○	○	◔	◐	◐	◐	◐	◐	◐	◐	●	◐	◐
Relativity Theory	○	○	○	○	○	○	○	○	◔	◔	◐	◐	◐
Fluid Behavior	○	○	○	○	○	○	○	○	◐	◐	◐	◐	◐
Science, Technology, & Mathematics													
Nature or Conceptions of Technology	○	◔	◔	◐	◐	◐	◐	◐	◐	◐	◐	◐	◐
Interactions of Science, Mathematics, & Technology													
Influence of Mathematics, Technology in Science	○	○	○	◔	◐	◐	◐	◐	◐	◐	◐	◐	◐
Applications of Science in Mathematics, Technology	○	○	○	◔	◐	◐	◐	◐	◐	◐	◐	◐	◐
Interactions of Science, Technology & Society													
Influence of Science, Technology on Society	○	○	○	◔	◐	◐	◐	◐	◐	◐	◐	◐	◐
Influence of Society on Science, Technology	○	○	○	◔	◐	◐	◐	◐	◐	◐	◐	◐	◐
History of Science & Technology	○	◔	○	◐	◐	◐	◐	◐	◐	◐	◐	◐	◐
Environmental & Resource Issues													
Pollution	○	◐	◐	◐	◐	◐	◐	●	●	●	●	●	◐
Conservation of Land, Water, & Sea Resources	◔	◐	◐	◐	◐	◐	◐	●	●	●	●	●	◐
Conservation of Material & Energy Resources	○	◐	◐	◐	◐	◐	◐	●	●	●	●	●	◐
World Population	○	◔	◔	◐	◐	◐	◐	◐	◐	◐	◐	◐	◐
Food Production, Storage	○	◔	◐	◐	◐	◐	◐	◐	◐	◐	◐	◐	◐
Effects of Natural Disasters	○	◐	◐	◐	◐	◐	◐	◐	◐	◐	◐	◐	◐
Nature of Science													
Nature of Scientific Knowledge	◔	◐	◐	◐	◐	◐	◐	◐	◐	◐	◐	◐	◐
The Scientific Enterprise	○	○	○	○	○	○	○	◐	◐	◐	◐	◐	◐
Science & Other Disciplines													
Science & Mathematics	○	◔	◔	◐	◐	◐	◐	◐	◐	◐	◐	◐	◐
Science & Other Disciplines	◔	◐	◐	◐	◐	◐	◐	◐	◐	◐	◐	◐	◐

Legend:

● at least 70% of the countries covered this topic at this grade level

◐ 10 to 70% of the countries covered this topic at this grade level

◔ at most 10% of the countries covered this topic at this grade level

○ no country covered this topic at this grade level

Figure 6.3

Proportions of Countries Focusing on Each Topic at Each Grade.

The same system as in the previous figure is used to portray topics and grades at which curricular attention was commonly focused. Far fewer common foci are seen than topics and grades with high common coverage.

Science Topics

Topic	P	1	2	3	4	5	6	7	8	9	10	11	12
Earth Sciences													
Earth Features													
Composition	○	○	○	◔	◔	◐	◐	◐	◐	◐	◐	◐	◐
Landforms	○	○	○	◐	◐	◐	◐	◐	◐	◐	◐	◐	◐
Bodies of Water	○	○	○	◐	◔	◐	◐	◐	◐	◐	◐	◐	◐
Atmosphere	○	○	◔	○	◔	◐	◐	◐	◐	◐	◐	◐	○
Rocks, Soil	○	◔	○	○	◔	◐	◐	◐	◐	◐	◐	◐	◐
Ice Forms	○	○	○	○	○	○	○	○	◐	◐	○	○	○
Earth Processess													
Weather & Climate	○	○	○	◐	◔	◐	◐	◐	◐	◐	◐	◐	○
Physical Cycles	○	○	○	○	◔	◐	◐	◐	◐	◐	◐	◐	○
Building & Breaking	○	○	○	○	○	◔	◐	◐	◐	◐	◐	◐	◐
Earth's History	○	○	○	○	○	○	○	◐	◐	◐	◐	◐	◐
Earth in the Universe													
Earth in the Solar System	○	○	○	◔	◐	◔	○	◐	◐	◐	◐	◐	◐
Planets in the Solar System	○	○	○	○	◐	◔	○	◐	◐	◐	◐	◐	◐
Beyond the Solar System	○	○	○	○	◔	○	○	○	◐	◐	◐	◐	◐
Evolution of the Universe	○	○	○	○	○	○	○	○	◐	◐	◐	◐	◐
Life Sciences													
Diversity, Organization, Structure of Living Things													
Plants, Fungi	○	○	◔	◐	◐	◐	◐	◐	◐	◐	◐	◐	◐
Animals	○	○	○	◐	◐	◐	◐	◐	◐	◐	◐	◐	◐
Other Organisms	○	○	○	○	◔	◐	◐	◐	◐	◐	◐	◐	◐
Organs, Tissues	○	○	○	◔	◔	◐	◐	◐	◐	◐	◐	◐	◐
Cells	○	○	○	○	○	◔	◐	◐	◐	◐	◐	◐	◐
Life Processes & Systems Enabling Life Functions													
Energy Handling	○	○	○	○	○	◐	◔	◐	◐	◐	◐	◐	◐
Sensing & Responding	○	○	○	◔	○	◔	◐	◐	◐	◐	◐	◐	◐
Biochemical Processes in Cells	○	○	○	○	○	○	○	◔	◐	◐	◐	◐	◐
Life Spirals, Genetic Continuity, Diversity													
Life Cycles	○	◔	○	◔	◐	◐	◐	◐	◐	◐	◐	◐	◐
Reproduction	○	○	○	○	○	◐	◐	◐	◐	◐	◐	◐	◐
Variation & Inheritance	○	○	○	○	○	○	◔	○	◐	◐	◐	◐	◐
Evolution, Speciation, Diversity	○	○	○	○	○	○	○	◐	◐	◐	◐	◐	◐
Biochemistry of Genetics	○	○	○	○	○	○	○	○	◐	◐	◐	◐	◐

Legend:

● at least 70% of the countries covered this topic at this grade level

◐ 10 to 70% of the countries covered this topic at this grade level

◔ at most 10% of the countries covered this topic at this grade level

○ no country covered this topic at this grade level

Figure 6.3 (continued)
Proportions of Countries Focusing on Each Topic at Each Grade.

Science Topics	P	1	2	3	4	5	6	7	8	9	10	11	12
Interactions of Living Things													
Biomes & Ecosystems	○	○	○	○	○	◐	◐	◐	◐	◐	◐	◐	◐
Habitats & Niches	○	○	○	○	○	◐	◐	◐	◐	◐	◐	◐	◐
Interdependence of Life	○	○	○	○	○	◐	◐	◐	◐	◐	◐	◐	◐
Animal Behavior	○	○	○	○	○	◐	◐	◐	◐	○	◐	◐	○
Human Biology & Health													
Nutrition	○	○	○	○	◐	◐	○	◐	◐	◐	◐	◐	◐
Disease	○	○	○	○	○	◐	◐	◐	◐	◐	◐	◐	◐
Physical Sciences													
Matter													
Classification of Matter	○	○	○	○	○	○	○	◐	◐	◐	◐	◐	○
Physical Properties	○	○	○	○	○	○	◐	◐	◐	◐	◐	◐	○
Chemical Properties	○	○	○	○	○	○	◐	◐	◐	◐	◐	◐	◐
Structure of Matter													
Atoms, Ions, Molecules	○	○	○	○	○	○	○	◐	◐	◐	◐	◐	◐
Macromolecules, Crystals	○	○	○	○	○	○	○	○	◐	◐	◐	◐	◐
Sub-atomic Particles	○	○	○	○	○	○	○	○	◐	◐	◐	◐	◐
Energy & Physical Processes													
Energy Types, Sources, Conversions	○	○	○	○	○	◐	◐	◐	◐	◐	◐	◐	◐
Heat & Temperature	○	○	○	○	◐	○	◐	◐	◐	◐	◐	◐	◐
Wave Phenomena	○	○	○	○	○	○	◐	○	○	○	◐	◐	◐
Sound & Vibration	○	○	○	○	○	○	◐	○	◐	◐	◐	◐	◐
Light	○	○	○	○	○	○	◐	◐	◐	◐	◐	◐	◐
Electricity	○	○	○	○	○	○	◐	◐	◐	◐	◐	◐	◐
Magnetism	○	○	○	○	○	○	◐	○	◐	◐	◐	◐	◐
Physical Transformations													
Physical Changes	○	○	○	○	○	○	○	◐	◐	◐	◐	◐	◐
Explanations of Physical Changes	○	○	○	○	○	○	○	◐	◐	◐	◐	◐	◐
Kinetic Theory	○	○	○	○	○	○	○	○	○	◐	◐	◐	◐
Quantum Theory & Fundamental Particles	○	○	○	○	○	○	○	○	○	○	◐	◐	◐
Chemical Transformations													
Chemical Changes	○	○	○	○	○	○	◐	◐	◐	◐	◐	◐	◐
Explanations of Chemical Changes	○	○	○	○	○	○	◐	◐	◐	◐	◐	◐	◐
Rate of Chemical Change & Equilibrium	○	○	○	○	○	○	○	○	◐	◐	◐	◐	◐
Energy & Chemical Change	○	○	○	○	○	○	○	○	◐	◐	◐	◐	◐

Science Topics

Science Topics	P	1	2	3	4	5	6	7	8	9	10	11	12
Organic & Biochemical Changes	○	○	○	○	○	○	○	○	◐	◐	◐	◐	◐
Nuclear Chemistry	○	○	○	○	○	○	○	○	◦	◐	◐	◐	◐
Electrochemistry	○	○	○	○	○	○	○	◦	◐	◐	◐	◐	◐
Forces & Motion													
Types of Forces	○	○	○	○	○	○	◦	◐	●	◐	◐	◐	◐
Time, Space, & Motion	○	○	○	○	○	○	◦	◐	◐	◐	◐	◐	◐
Dynamics of Motion	○	○	○	○	○	○	◦	◐	◐	◐	◐	◐	◐
Relativity Theory	○	○	○	○	○	○	○	○	○	◦	◦	◐	◐
Fluid Behavior	○	○	○	○	○	○	○	◦	◦	◐	◐	○	◐
Science, Technology, & Mathematics													
Nature or Conceptions of Technology	○	◦	◦	◦	◦	◦	◐	○	◐	○	○	◐	◐
Interactions of Science, Mathematics, & Technology													
Influence of Mathematics, Technology in Science	○	○	○	○	○	○	○	◦	○	○	○	○	○
Applications of Science in Mathematics, Technology	○	○	○	○	○	○	○	◦	○	○	○	◐	◐
Interactions of Science, Technology & Society													
Influence of Science, Technology on Society	○	○	○	○	○	○	○	○	○	○	○	○	○
Influence of Society on Science, Technology	○	○	○	○	○	○	○	○	○	○	○	◐	◐
History of Science & Technology	○	◦	◦	◦	◦	◦	○	◦	◐	○	◐	◐	◐
Environmental & Resource Issues													
Pollution	○	○	○	○	○	○	○	●	◐	◐	●	◐	◐
Conservation of Land, Water, & Sea Resources	○	○	○	○	○	○	○	◐	◐	◐	◐	◐	◐
Conservation of Material & Energy Resources	○	○	○	○	○	○	○	◐	◐	◐	◐	◐	◐
World Population	○	○	○	○	○	○	○	◐	◐	◐	○	◐	◐
Food Production, Storage	○	○	○	○	○	○	○	○	◐	◐	○	◐	○
Effects of Natural Disasters	○	○	○	○	○	○	○	○	○	○	◦	◐	◐
Nature of Science													
Nature of Scientific Knowledge	○	◦	◦	◦	◦	◦	◦	◦	◐	○	◐	◐	○
The Scientific Enterprise	○	○	○	○	○	○	○	○	○	○	○	○	○
Science & Other Disciplines													
Science & Mathematics	○	○	○	○	○	◦	◦	◦	◦	○	○	○	○
Science & Other Disciplines	○	◦	◦	◦	○	◦	◦	○	○	○	○	○	○

Legend:

● at least 70% of the countries focused on this topic at this grade level

◐ 10 to 70% of the countries focused on this topic at this grade level

◦ at most 10% of the countries focused on this topic at this grade level

○ no country focused on this topic at this grade level

Unless the items used to measure attainments are largely insensitive to curricular differences, careful use of the curricular context of items must be an important part of interpreting achievement results. To achieve little in the face of limited relevant opportunity to learn is uninformative.

Commonly Intended Topics at Focal Grades

One of the strengths of the TIMSS' curriculum analysis is that such expert opinion can be supplemented by more direct analyses of key curricular documents. If one re-examined the percentage of countries that indicated particular science topics were to be covered — but using direct analysis of curriculum guides rather than expert judgments — would the portrait of commonly intended topics change in significant ways? The data to answer this question are available, but a tradeoff is essential. Curriculum guides and textbooks sampled in the TIMSS curriculum analysis focused on key grades relevant to the TIMSS achievement testing rather than on the full range of schooling. The expert data reviewed earlier was less directly based on curricular documents but included all of the grades in the flow of science curricula across the years of schooling. If attention is limited here to particular grades — in this case, the upper of the two grades for which achievement testing is to take place for TIMSS Populations 1 and 2 — a more detailed portrait emerges of common intentions at these critical points in schooling.

Table 6.1 presents data from science curriculum guides on the most commonly intended science framework topics. These data are only for the upper grades of TIMSS Populations 1 and 2 — that is, the higher of the two grades containing the majority of 9-year-old and 13-year-old students, respectively, in each country. The table lists the science framework topics found in the curriculum guides for the indicated grade levels of at least 70 percent of the TIMSS countries — a level comparable to high common coverage shown in figures 6.2 and 6.3. Further information is provided by the arrangement of these lists. Topics found widely *only* in curriculum guides, without mention and support in science textbooks for the grade level in question, are in the pair of lists at the top. Topics found in curriculum guides and supported by wide inclusion in the appropriate textbooks are in the middle pair of lists. Some topics were included widely in textbooks but were not in the corresponding curriculum guides. These are in the bottom pair of lists. Finally, topics that were found in the appropriate textbooks of at least 70 percent of the countries in more than 6 percent[1] of the blocks of those countries were considered as being emphasized in the textbooks. These topics are indicated with an asterisk in table 6.1.

As suggested by the curricular flow data in chapter 5, the commonly intended science curriculum at the upper grade of TIMSS Population 1 was essentially a curriculum of taxonomy in earth and life sciences, with some elements of physical science. 'Plants, fungi' and 'animals' were included widely in the curriculum guides and emphasized in textbooks. 'Weather and climate,' 'organs and tissues,' 'earth in the solar system,' 'interdependence of life,' and some topics in the physical sciences also were included in the appropriate curriculum guides of more than 70 percent of the countries and supported — but not emphasized — by textbooks.

Table 6.1
Commonly Intended Science Topics.

Here are lists of the most commonly intended (at least 70 percent of countries) science topics at the upper grades of Populations 1 and 2 based on an analysis of curriculum guides and textbooks. Some topics were common only in curriculum guides, others only in textbooks, and still others in both. A few were not only included but were also emphasized in textbooks.

UPPER GRADE OF POPULATION 1	UPPER GRADE OF POPULATION 2
Curriculum Guides (Not in Textbooks):	
Earth Sciences	**Physical Sciences**
Earth Features	*Structure of Matter*
Bodies of Water	*Subatomic Particles*
Life Sciences	*Energy and Physical Processes*
Human Biology and Health	*Magnetism*
Environmental and Resource Issues	**Life Sciences**
Pollution	*Diversity, Organization, Structure of Living Things*
Material and Energy Resources Conservation	*Other Organisms*
World Population	*Cells*
Food Production, Storage	*Life Spirals, Genetic Continuity, Diversity*
	Evolution, Speciation, Diversity
	Interactions of Living Things
	Biomes and Ecosystems
	Habitats and Niches
	Animal and Plant Behavior
	Human Biology and Health
	Environmental and Resource Issues
	World Population
	Food Production, Storage
	Effects of Natural Disasters
	Nature of Science
	Nature of Scientific Knowledge
Curriculum Guides (Included in Textbooks):	
Earth Sciences	**Physical Sciences**
Earth Processes	*Matter*
Weather and Climate	*Classification of Matter*
Earth in the Universe	*Physical Properties*
Earth in the Solar System	*Chemical Properties*
Life Sciences	*Structure of Matter*
Diversity, Organization, Structure of Living Things	*Atoms, Ions, Molecules*
*Plants, Fungi	*Energy and Physical Processes*
*Animals	*Energy Types, Sources, Conversions*
*Organs, Tissues	*Heat and Temperature*
Interactions of Living Things	*Light*
Interdependence of Life	*Life Processes & Systems Enabling Life Functions*

* **Note:** Topic emphasized in textbooks.

Table 6.1 (continued)
Commonly Intended Science Topics.

Physical Sciences	Energy Handling	Electricity
Matter	Sensing and Responding	*Chemical Transformations*
Physical Properties	*Life Spirals, Genetic Continuity, Diversity*	Explanations of Chemical Changes
Energy and Physical Processes	Life Cycles	*Forces and Motion*
Energy Types, Sources, Conservation	Reproduction	Types of Forces
Environmental and Resource Issues	*Interactions of Living Things*	**Science, Technology, and Mathematics**
Land, Water, and Sea Resources	Interdependence of Life	*Interactions of Science, Mathematics, and Technology*
	Human Biology and Health	Science Applications in Mathematics, Technology
	Disease	*Interactions of Science, Technology, and Society*
		Influence of Science Technology on Society
		Environmental and Resource Issues
		Pollution
		Conservation of Land, Water, and Sea Resources
		Conservation of Material and Energy Resources

Exclusively in Textbooks:

Earth Sciences
Earth Features
Rock, Soil

Physical Sciences
Physical Transformations
Explanations of Physical Changes
Forces and Motion
Time, Space, and Motion

History of Science and Technology

* *Note:* Topic emphasized in textbooks.

There was a group of topics in the curriculum guides for the upper grade of Population 1 in more than 70 percent of the TIMSS countries that was not supported by corresponding materials in the textbooks. This group included materials on 'bodies of water' in the earth sciences and 'human biology and health' in the life sciences. It is noteworthy that a variety of topics in 'environmental and resource issues related to science' — regarded by many TIMSS experts as an important element of science education reform in their countries (see chapter 2) — were found in more than 70 percent of the curriculum guides but were not supported in the textbooks. Curricular intention appears to be most clearly expressed by curriculum guides — with a secondary, reflective role played by textbooks in many countries. Textbooks also play differing roles in classroom instruction in different countries, ranging from being a virtually invisible support used primarily by the teacher in some countries, to being a de facto determinant of the curriculum (at times in spite of official statements) in others.

Care must be taken in drawing conclusions from the existence of commonly intended topics in curriculum guides that are not also supported in textbooks. However, the criterion of inclusion in curriculum guides for 70 percent of the countries makes it unlikely that these topics occurred in the guides only of those countries that made no significant use of textbooks. There clearly was some disagreement in the intentions of those that envisioned these science curricula and set their aims, and those who supported the implementation of the curricula by writing science textbooks. At the very least, these data suggest that student attainments in these areas must be carefully examined to help determine the relative impact of intentions articulated by curriculum guides and intentions embodied — if not articulated in specific objectives — in textbooks supporting instruction. Additionally, it must be noted that in many countries there is an inevitable lag between publication of new curriculum guides and the development and marketing of new textbooks to support them.

The commonly intended science contents at the upper grade of Population 2 were more diverse than the corresponding set for Population 1, perhaps reflecting the predominant "additive" pattern of curriculum in the sciences. For this grade, the commonly intended topics included all those commonly intended for the upper grade of Population 1 except two: 'bodies of water' and 'earth in the solar system.' Physical sciences topics emerge as more heavily emphasized in this curriculum portrait — including new topics (relative to the upper grade of Population 1) in the areas of the 'structure of matter,' 'energy and physical processes,' and chemistry. The only topic specified widely in curriculum guides and emphasized — rather than merely supported — by textbooks was that of 'organs, tissues.'[2]

With the upper grade of Population 2, another category of topics emerged. Several topics appeared exclusively in textbooks and were not widely supported in curriculum guides. This suggests that the curricular intentions implicit in the inclusion of materials in textbooks were not always in full agreement with the explicit intentions of curriculum guides — and that the implicit aims of textbooks were somewhat more inclusive than the aims of guides. Given different practices in textbook use — and widespread practices of teacher selectivity of textbook content covered — and pending direct data from teachers on content emphases — no firm conclusions can be drawn about this material in textbooks not supported in curriculum guides. The direct teacher data needed for further understanding is being collected as part of TIMSS.

In the upper grade of Population 2 there are also more topics called for in guides that are not supported in textbooks. These include 'human biology and health,' which also was called for in Population 1 curriculum guides and not supported in textbooks.

In addition, it is worth noting that aside from the traditional disciplines of life, earth, and physical sciences, another area is included in curriculum guides and supported (although not emphasized) in textbooks. These are topics dealing with science applications and technology.

One feature of the data in table 6.1 is that for the upper grades of both Populations 1 and 2 only a few science framework topics were widely included in curriculum guides and were both supported and emphasized in textbooks. Having moved from expert data to document-based data, it is possible to estimate fairly precisely what emphasis means in these cases. Two Population 1 topics were emphasized — 'plants, fungi' and 'animals.' One Population 2 topic was emphasized: 'organs, tissues.' These topics appeared in more than 70 percent of the countries' curriculum guides and more than 70 percent of the countries' textbooks. Further, at least 6 percent of the textbook blocks in each of these 70 percent of the countries were coded to indicate treatment of that content.

This indicates how widespread was the inclusion of these topics. But how extensive was the textbook emphasis? One emphasis measure is the average percentage of total textbook blocks devoted to each of these topics across the countries that included these contents in the appropriate grade-level textbook. This measure yields the relative amount of space on average devoted to each of the topics. Since the measure is a proportion of blocks, it compensates for differences among coders in block sizes, linguistic densities, textbook sizes, etc; that is, it provides a self-adjusting measure of relative emphasis.

On average, only a bit more than a quarter of the Population 1 upper grade textbooks' blocks were devoted to the combination of the emphasized topics. Thus, although a significant portion of the Population 1 curriculum appears to have been plant and animal taxonomy, this topic accounts for only about a quarter of the curriculum. Of course, the actual proportion of this material used, as well as the extent to which this material was supplemented, can only be precisely determined by direct teacher data. Even so, these data are strongly suggestive of the importance of these central, commonly intended topics and help to further characterize the Population 1 curriculum. The data for the upper grade of Population 2 are much less significant — the single emphasized topic ('organs, tissues') accounted for only about 6 percent of a textbook's blocks on average.

These data present a noteworthy contrast with similar data from mathematics curricular documents. In *science,* the two commonly intended topics receiving most emphasis in textbooks intended for the upper grade of Population 1 take up about one-quarter of textbook space.

In *mathematics,* approximately 60 percent of space in textbooks are devoted to the most emphasized commonly intended topics. A similar phenomenon occurs for the upper grade of Population 2 — whereas only 6 percent of science textbook space is devoted to commonly intended emphasized topics, these type of topics take up about 20 percent of mathematics textbooks.

For the targeted grades in TIMSS populations, a wealth of information now characterizes more precisely the science content of the curriculum. Data are available on which topics were widely mentioned in curriculum guides and which were widely supported by inclusion in science textbooks. By including expert data on the flow of curriculum, the portrait can be extended to include whether a topic had been treated in the curriculum prior to a targeted grade or was intended for later treatment.

The task of coordinating these data and forming a more complete overall portrait of the particular topic as milestone content (or not) in a particular country is daunting. One attempt to combine these varying data for the upper grade of Population 2 is given in appendix I, figure I6.4. Data for a representative selection of countries and topics are shown in figure 6.4.

In this figure, each column represents a country and each row a topic. Each cell is thus a combination of topic and country — a visual summary of the treatment of that topic in the upper grade of Population 2. Each cell contains potentially four symbols: an arrow pointing to the left, an arrow pointing to the right, and, between them, either a dash or cross. Any of these symbols can be absent in any case. The arrow to the left indicates that curricular flow data suggested that this topic was intended to be covered by this country in at least one grade prior to the upper grade of Population 2. An arrow to the right similarly indicates that the data suggested that it was intended that this topic be covered in at least one grade after the upper grade of Population 2. The presence of both arrows indicates that the target grade was centered in 3 or more years of intended coverage.

In the center position, a dash indicates that the topic for that country appeared in its curriculum guide as intended for coverage at the upper grade of Population 2, but did not appear in the corresponding textbooks. A cross indicates that the topic both appeared in the curriculum guide and was supported by a relevant textbook for the upper grade of Population 2 in that country. The data presented in the center position comes from the document analyses; the arrows from the flow data are supplied by expert opinion.

Almost every possible combination of the four symbols is found in some cell. For some countries, coverage at this Population 2 target grade came at the end of previous coverage; for others it came at the beginning of continued coverage. In only two cases was the coverage restricted to this single upper grade of Population 2: for 'physical properties of matter' and 'chemical properties of matter' in Hungary. The figure's pattern is one of considerable diversity in terms of textbook support, prior coverage, and intended future coverage. Countries varied from topic to topic; topics varied among countries. The data again suggest how difficult it

Figure 6.4

Coverage of Commonly Intended Topics Over Time.

The cross-section of commonly intended topics at the upper grade of Population 2 is here supplemented by the curricular flow data to show whether a topic was widely covered in guides and/or textbooks, was intended to be covered before, or was intended to be covered later. This figure presents data for a sample of countries and topics.

Science Topics

	Australia	Belgium (FI)[1]	Hungary	Israel	Latvia	Netherlands
Bodies of Water	← →	← - →			← - →	← + →
Animals	← + →	← - →	← +	← + →	← + →	← - →
Organs, Tissues	← + →	← →	+ →	← + →	← + →	←
Evolution, Speciation, Diversity	+ →	- →	+ →	← + →	- →	- →
Disease	← - →	← - →	+ →	← + →	←	← →
Physical Properties of Matter	← + →	← - →	+	← + →	+ →	→
Chemical Properties of Matter	+ →	→	+	← + →	+ →	+ →
Atoms, Ions, Molecules	+ →		← →	← + →	+ →	
Energy Types, Sources, Conversions	← + →	←	← →	← - →	+ →	←
Light	← + →	- →	+ →	← + →	+ →	→
Electricity	+ →	- →	+ →	← + →	- →	→
Explanations of Chemical Changes	→	- →	→	→	- →	→
Organic & Biochemical Changes	→	- →	→	→	- →	+ →
Applications of Science in Mathematics, Technology	+ →	← - →	+ →	+	+ →	← + →

Science Topics

	Philippines	Portugal	Russian Federation	Sweden[2]
Bodies of Water	← →	← - →	← +	← + →
Animals	← + →	← + →	← +	← +
Organs, Tissues	← + →	← + →	← + →	← + →
Evolution, Speciation, Diversity	→	←	← - →	← + →
Disease	←	←	← →	← + →
Physical Properties of Matter	← + →	←		← +
Chemical Properties of Matter	+ →	←		- →
Atoms, Ions, Molecules	← →	+ →	← + →	← +
Energy Types, Sources, Conversions	← + →	← + →	← + →	← - →
Light	← - →	← + →	+ →	← →
Electricity	← - →	+ →	+ →	← +
Explanations of Chemical Changes	→	+ →	- →	
Organic & Biochemical Changes	→	← + →	+ →	
Applications of Science in Mathematics, Technology	← →	← + →	← →	←

[1] The National Research Coordinators of Belgium have collected data only from curriculum guides. Due to the great level of detail of the guides, and their extensive use, data from these are compared in this display with the textbook data supplied from all other countries.

[2] In Sweden, curriculum objectives apply to groups of grades, not individual grades. Caution is suggested when interpreting these data. Though chemistry textbooks do exist for this level, none were coded.

Legend:
← Intended for coverage in grades prior to the upper grade of Population 2.
→ Intended for coverage in grades after the upper grade of Population 2.
+ In both the curriculum guide and the textbook at the upper grade of Population 2.
- Present in the curriculum guide at the upper grade of Population 2.

was to specify a reasonable set of common achievement items to measure science attainments in this context of diversity in which virtually every comparison must be carefully interpreted. Some topics were clearly inappropriate for some countries — if all coverage was intended for the future — although that was rare. Probably the highest level of appropriateness was in the case of a cross surrounded by both a right and left arrow — for this situation the topic involved was in mid-sequence instructionally, was targeted in the curriculum guide as a curricular aim, and had its implementation supported by materials from an appropriate textbook. Other patterns indicate lower levels of appropriateness.

Attention now turns to Population 3. While achievement testing of this population is for a sample of all students at the end of secondary school on quantitative and scientific literacy, a subpopulation has been defined as those students currently taking advanced physics. Consideration here is restricted to this sub-sample of physics "specialists," since they were the only students at the end of secondary school for which curricular documents were sampled.

With this restriction in mind, a portrait can be made of the topics commonly intended for these physics specialists at the end of secondary school. The list of those topics included in at least 70 percent of the appropriate science curriculum guides is found in table 6.2. As before, these lists are arranged by whether the topics were in curriculum guides only, in curriculum guides and supported by textbooks, or only supported in textbooks. Emphasis in textbook support is again indicated by an asterisk (covered in at least 6 percent of textbook blocks).

The specialist curriculum — unsurprisingly in view of its definition — was primarily one of physical sciences with some topics in science applications and technology (called for in guides, not supported by textbooks). What is perhaps more surprising is the preponderance of topics that would appear to indicate a curriculum mostly devoted to 19th century physics.

Six topics, five in the area of energy and physical processes and one in the area of forces and motion, were not only supported, but also emphasized in textbooks. A large portion (approximately 70 percent) of textbooks are commonly devoted to these emphasized commonly intended topics — almost 20 percent of the books are devoted to the topic of 'electricity' alone. The remaining five topics are commonly present in approximately 10 percent of the blocks each (somewhat less for 'energy types, sources, conversions'). The topic of 'energy types, sources, conversions' is also commonly intended in the upper grades of Populations 1 and 2.

The list of topics in 70 percent of textbooks, but not called for in 70 percent of curriculum guides, is noteworthy: 'kinetic-molecular theory' and 'relativity theory.'

Table 6.2

Commonly Intended Topics for Physics Specialists.

Here are the common topics for physics specialists at the end of secondary school. These topics were in the curriculum guides and/or textbooks of at least 70 percent of the countries; those indicated with an asterisk were also emphasized in textbooks.

POPULATION 3 SPECIALISTS

Curriculum Guides (Not in Textbooks):

Atoms, Ions, Molecules
Macromolecules, Crystals
Heat and Temperature
Nuclear Chemistry
Science Applications in Mathematics, Technology

Curriculum Guides (Included in Textbooks):

Subatomic Particles
* Energy Types, Sources, Conversions
* Wave Phenomena
Sound and Vibration
* Light
* Electricity
* Magnetism
Quantum Theory, Fundamental Particles
Types of Forces
* Time, Space, and Motion
Dynamics of Motion

Exclusively in Textbooks:

Kinetic-Molecular Theory
Relativity Theory

---

* ***Note:*** Topic emphasized in textbooks.

NOTES

1   Six percent was determined empirically as a cutoff level that separated textbooks that devoted considerable space to a topic — here considered as emphasizing it — from those that gave only some (comparatively little) space to covering the topic.

2   This category of the science framework includes both the structure and function of tissues, organs, and organ systems (for example, digestion, excretion, circulation).

Chapter 7

EMPHASIZED CURRICULUM TOPICS

Analysis of the curricular visions and aims of the TIMSS countries revealed a number of commonly intended topics at several grade levels. Underlying these similarities are myriad differences in coverage patterns and emphases. How many topics did a country address in a year? Did these topics receive equal emphasis? Was coverage of a particular topic concentrated over one school year or diffused over several years? How much emphasis did textbooks give to these topics? No single indicator or analytic method can address these questions. Consequently, we used several measures, including the following ones listed below:

- Topics covered in both curriculum guides and textbooks;

- Number of topics covered during a school year;

- Share of curricular attention devoted to each topic; and

- Percentage of textbook blocks devoted to specific topics.

Each of these indicators yields estimates of at least one facet of emphasis or intensity in curricular attention. Each measure has its own strengths and weaknesses.

Emphasis as Agreement of Curriculum Guides and Textbooks

How can one determine what topics countries emphasized in their science curricula? One measure is to see what was covered by their curricular materials. Those topics that both curriculum guides and textbooks agreed should be covered at a particular grade are special objects of curricular attention; they probably received more emphasis than topics without such agreement.

Data on these agreements could only be considered on a by-grade basis; in the analysis they were limited to those grades for which data from both curriculum guides and textbooks were available. Table 7.1 presents these data for the upper grade of Population 2 for all countries providing the needed data and for each of the science framework topics.

A circle indicates that the topic was present in the relevant curriculum guide or guides for the country. A cross indicates that the topic was present in the relevant textbook or textbooks

Table 7.1
Topics Present in Both Curriculum Guides and Textbooks.

Topics present in both the curriculum guides and textbooks for a given grade level in a country probably received more emphasis than those not present in both. The data for all topics and all countries are shown here for the upper grade of Population 2. Few topics received unanimous emphasis in all countries, but many were widely emphasized.

Science Topics	Australia	Austria	Canada	Colombia	Czech Republic	Denmark[a]	Dominican Republic	France	Germany	Greece	Hong Kong	Hungary	Iceland	Ireland
Earth Features														
Composition	⊕	⊕	⊕	⊕	⊕			⊕	+	+	+	+	+	
Landforms	⊕	⊕	⊕	⊕	⊕			⊕	+	+	+	+	⊕	
Bodies of Water	+	○	⊕	⊕	⊕	○		⊕	+	+	+	+	⊕	⊕
Atmosphere	⊕		⊕	⊕	⊕	○		○	+	+	⊕	+	+	⊕
Rocks, Soil	⊕	⊕	⊕	⊕	⊕			⊕	+	⊕	+	+	+	⊕
Ice Forms	+		⊕	⊕	⊕			○	+	+	+		⊕	
Earth Processes														
Weather & Climate	⊕	⊕	⊕	⊕	⊕	○		○	+	+	+		⊕	⊕
Physical Cycles	⊕	○	⊕	⊕	+			⊕	+	+	+	+	⊕	⊕
Building & Breaking	⊕	+	⊕	⊕	+			⊕	+	+	+	+	⊕	+
History	⊕	⊕	⊕	⊕	⊕			⊕	+	+	+	+	⊕	○
Earth in the Universe														
Earth in the Solar System	⊕	+	⊕	⊕	+	○	○	○	+			+	⊕	⊕
Planets in the Solar System	○	⊕	⊕	⊕	+	○	○	○				+	⊕	⊕
Beyond the Solar System	○		+	⊕		○	○	⊕				+	⊕	⊕
Evolution of the Universe	○	+	+	⊕		○	○	○				+	+	
Diversity, Organization, Structure of Living Things														
Plants, Fungi	⊕	⊕	⊕	⊕	⊕		⊕	+	○	+		⊕	⊕	⊕
Animals	⊕	+	⊕	⊕	⊕		⊕		○	+		⊕	⊕	⊕
Other Organisms	⊕	+	⊕	⊕			⊕		○	⊕		⊕	⊕	⊕
Organs, Tissues	⊕		⊕	⊕			⊕		○	⊕	⊕	⊕	⊕	⊕
Cells	⊕		⊕	⊕		○			○	+		⊕	⊕	⊕
Life Processes & Systems Enabling Life Functions														
Energy Handling	⊕	⊕	⊕	⊕	⊕	○	+	+	○	⊕	⊕	⊕	⊕	⊕
Sensing & Responding	⊕	+	⊕	⊕		⊕	+	+	○	⊕	⊕	⊕	⊕	⊕
Biochemical Processes in Cells	⊕		⊕	⊕				+	○	+		⊕		⊕
Life Spirals, Genetic Continuity, Diversity														
Life Cycles	⊕	⊕	⊕	⊕		○	○	○	○			⊕	⊕	⊕
Reproduction	⊕	+	⊕	⊕			⊕	⊕	○	⊕		⊕	⊕	⊕
Variation & Inheritance	+	⊕	⊕	⊕			⊕			⊕		⊕	○	+
Evolution, Speciation, Diversity	⊕	⊕	⊕	⊕	⊕		○	○	○			⊕	⊕	+
Biochemistry of Genetics	+		⊕	⊕			○					+		+

a　Refer to footnotes to figure 16.4

b　The country has requested that this topic be noted as absent in curriculum guides despite the inconsistency with the original coded data.

c　The country has requested that this topic be noted as absent in textbooks despite the inconsistency with the original coded data.

d　The country has requested that this topic be noted as present in curriculum guides despite the inconsistency with the original coded data.

e　New Zealand data include old and new curricula. Caution must be exercised in interpretation since the new curriculum spans all three focal populations, whereas the old curriculum states by-grade objectives that have been separated for each population.

f　In Scotland, textbooks are used infrequently in science education. Caution is suggested when comparing contents of guides to the of textbooks.

g　The Swedish curriculum guide for science states objectives for all grades from 1 through 9 inclusive, without detailing by-grade objectives. Thus, in this table, curriculum guide data are objectives for grades 1 through 9, while textbook data are only for grade 7 (the upper population of Population 2).

Missing Data: Argentina (no units of appropriate type), Belgium, Bulgaria, People's Republic of China, Iran (no official guides exist), Latvia, Lithuania, Thailand, and Tunisia.

Israel	Japan	Korea	Latvia	Mexico	Netherlands	New Zealand[e]	Norway	Philippines	Portugal	Romania	Russian Federation	Scotland[f]	Singapore	Slovak Republic	Slovenia	South Africa	Spain	Sweden[g]	Switzerland	USA
○		⊕	⊕	+	+	○	⊕	+	○	+		⊕		⊕	○	+	+	+	+	⊕
		⊕	○	⊕	⊕	○	⊕	+	○	⊕		○		⊕	+○	⊕	+	⊕	⊕	⊕
		+	⊕	⊕	⊕	○	⊕	+	+	⊕		○		⊕	+○	+	⊕	⊕	⊕	⊕
		⊕	○	⊕	⊕	○	⊕	+	+○	⊕		○		⊕	+○	+	+	⊕	⊕	⊕
		+			⊕	○	○	+	○	+		○		⊕		+	+	+	+	⊕
	[bc]	⊕	⊕	⊕	⊕	○	⊕	⊕	+	⊕		⊕	+	⊕	⊕	⊕	⊕	○	⊕	⊕
	+	⊕	○	+	+	⊕	⊕	⊕	○			⊕		⊕	○	+	⊕	○	⊕	⊕
○		⊕	○	⊕	○	○	⊕	⊕	○	+		⊕		⊕	○		⊕		⊕	⊕
		⊕	○	○		○	⊕	⊕	○	+		⊕		⊕	⊕	⊕	⊕		⊕	⊕
		⊕	⊕	+	⊕	⊕	+	⊕	⊕			⊕		+	⊕	⊕	⊕	⊕	+	⊕
		○	⊕		⊕	⊕	+	⊕	⊕			⊕		+	⊕	⊕	⊕	⊕	+	⊕
		⊕	⊕		⊕	⊕	+	⊕	+			○			○	⊕	⊕	+	+	⊕
		○	⊕		○	○	+	⊕	⊕			○			○	○	+	⊕	+	⊕
⊕		+	○	⊕	○	⊕	⊕	⊕	⊕	+	○	⊕	○	⊕	⊕	⊕	⊕	⊕	⊕	⊕
⊕	⊕	+	○	⊕	⊕	⊕	⊕	⊕	⊕	+	⊕	⊕	○○	⊕	⊕	⊕	⊕	⊕	⊕	⊕
○		+	○	⊕	⊕	⊕	⊕	○	⊕	+	○	⊕	○○	⊕	⊕	⊕	⊕	⊕	⊕	⊕
⊕	⊕	⊕[d]	○	⊕	+	⊕	⊕	⊕	⊕	+	⊕	○		⊕	⊕	⊕	⊕	+	⊕	⊕
		+	○	⊕		⊕	⊕	⊕	⊕	+	○	○			⊕	⊕	⊕		⊕	⊕
⊕	⊕	⊕	○	⊕	⊕	⊕	⊕	+	⊕		⊕	⊕	⊕	⊕	⊕	⊕	⊕	⊕	⊕	⊕
⊕	⊕	⊕	⊕	⊕	⊕	⊕	⊕	+	⊕	⊕	⊕	⊕	○	⊕	⊕		○	⊕	⊕	⊕
○		⊕	○	⊕	⊕	⊕	⊕	+	⊕		○	○	⊕		+	○	⊕	+	⊕	⊕
+	○[d]		○	⊕	⊕	⊕	⊕	+	⊕	+	⊕	○	⊕		⊕	⊕	⊕		⊕	⊕
⊕	○[d]	+	○	⊕	+	○	⊕	+	⊕	⊕	○	○			⊕	⊕	⊕	⊕	⊕	⊕
⊕	○[d]		○	⊕	⊕	○	⊕		+	+	⊕	○		⊕	⊕	○	⊕		⊕	⊕
○			○	⊕		○	⊕		+		⊕	○			+	⊕	○	⊕		⊕

<table>
<tr><td>Legend:</td><td>+ - Present in Textbooks
○ - Present in Curriculum Guides</td></tr>
</table>

Table 7.1 (continued)
Topics Present in Both Curriculum Guides and Textbooks.

Science Topics	Australia	Austria	Canada	Colombia	Czech Republic	Denmark	Dominican Republic	France	Germany	Greece	Hong Kong	Hungary	Iceland	Ireland
Interaction of Living Things														
Biomes & Ecosystems	⊕	⊕	⊕	⊕			○		○	+	+	○	⊕	⊕
Habitats & Niches	⊕	○	⊕	⊕			○		○		+	+	⊕	⊕
Interdependence of Life	⊕	⊕	⊕	⊕	○		⊕	○	○		+	+	⊕	⊕
Animal Behavior	⊕		⊕				○		○		+	⊕	⊕	⊕
Human Biology and Health														
Human Biology & Health	⊕	⊕	⊕		○	⊕	+	⊕	○	⊕		+	⊕	⊕
Nutrition	⊕	⊕	+	+						⊕		⊕	⊕	⊕
Disease	⊕	⊕	⊕	⊕			+	⊕	○	+		⊕	○	⊕
Matter														
Classification of Matter	⊕	⊕	⊕	⊕	⊕	○	+	⊕	+	⊕		⊕	○	⊕
Physical Properties	⊕	⊕	⊕	⊕	+	⊕	⊕	⊕	+	⊕	○	⊕	⊕	⊕
Chemical Properties	⊕	⊕	⊕	⊕	⊕	○	⊕	⊕	+	⊕		⊕		⊕
Structure of Matter														
Atoms, Ions, Molecules	⊕	⊕	⊕	+	○	○	○	○	+	⊕		⊕	⊕	⊕
Macromolecules, Crystals	⊕	⊕	⊕	○	+	○	○	○		⊕		⊕	⊕	+
Sub-atomic Particles	⊕	⊕	+		+	○	○	○		⊕		⊕	⊕	⊕
Energy and Physical Processes														
Energy Types, Sources, Conversions	⊕	⊕	⊕	⊕	⊕	⊕	⊕		⊕	⊕	⊕	⊕	○	⊕
Heat & Temperature	+	⊕	⊕	⊕	⊕	⊕	⊕		+	⊕	⊕	○	○	○
Wave Phenomena	+	+	+	○	⊕	+	○	⊕				○	○	⊕
Sound & Vibration	⊕	⊕	+	⊕	+	+	○	⊕			⊕	⊕	○	⊕
Light	⊕	⊕	+	⊕	+	○	⊕	⊕	⊕	+	○	⊕	○	⊕
Electricity	⊕	⊕	+	⊕	⊕	○	⊕	○			⊕	⊕	⊕	⊕
Magnetism	○	⊕	+	○	⊕	○	⊕				⊕	⊕	⊕	⊕
Physical Transformations														
Physical Changes	⊕	⊕	⊕	⊕				⊕	⊕	⊕		⊕	⊕	⊕
Explanations of Physical Changes	⊕	⊕	⊕	+	+	○		+		+		○	⊕	○
Kinetic-Molecular Theory	○	⊕	⊕							⊕		○	○	+
Quantum Theory, Fundamental Particles	○	⊕	+			○							○	
Chemical Transformations														
Chemical Changes	⊕	⊕	+		⊕	○	○	⊕	+	⊕	⊕	+	⊕	⊕
Explanation of Chemical Changes	+	⊕		+	⊕	○	⊕	+	+			+		⊕
Rate of Chemical Change & Equilibrium	+	+		+	⊕	○	○	+	○					
Energy & Chemical Change	⊕	⊕		+	⊕	○	○		+	+		+	○	○
Organic & Biochemical Changes	+	⊕	+	+	⊕	○	○					+		
Nuclear Chemistry	+	⊕	+	+	○	○	○	○				+	○	○
Electrochemistry	+	⊕	+	+	⊕	○	○	○		+	⊕			⊕

a Refer to footnotes to figure 16.4

b The country has requested that this topic be noted as absent in curriculum guides despite the inconsistency with the original coded data.

c The country has requested that this topic be noted as absent in textbooks despite the inconsistency with the original coded data.

d The country has requested that this topic be noted as present in curriculum guides despite the inconsistency with the original coded data.

e New Zealand data include old and new curricula. Caution must be exercised in interpretation since the new curriculum spans all three focal populations, whereas the old curriculum states by-grade objectives that have been separated for each population.

f In Scotland, textbooks are used infrequently in science education. Caution is suggested when comparing contents of guides to the of textbooks.

g The Swedish curriculum guide for science states objectives for all grades from 1 through 9 inclusive, without detailing by-grade objectives. Thus, in this table, curriculum guide data are objectives for grades 1 through 9, while textbook data are only for grade 7 (the upper population of Population 2).

Missing Data: Argentina (no units of appropriate type), Belgium, Bulgaria, People's Republic of China, Iran (no official guides exist), Latvia, Lithuania, Thailand, and Tunisia.

Israel	Japan	Korea	Latvia	Mexico	Netherlands	New Zealand[a]	Norway	Philippines	Portugal	Romania	Russian Federation	Scotland[f]	Singapore	Slovak Republic	Slovenia	South Africa	Spain	Sweden[g]	Switzerland	USA
⊕			○	⊕	⊕	○	⊕	+	○	⊕	⊕	○	⊕		○	⊕	⊕	○	⊕	⊕
⊕			○	⊕	⊕	⊕	⊕	+	○	⊕	⊕	○	⊕		○	⊕	⊕	○	⊕	⊕
⊕	+[b]		○	⊕	⊕	⊕	⊕	+	○	⊕	⊕	○	⊕	○	⊕	⊕	⊕	+	⊕	⊕
○		+	⊕	○	⊕	⊕	⊕		⊕	⊕	○	+	○	○	⊕			⊕	⊕	⊕
+				⊕	⊕	⊕	⊕	+	⊕	⊕		⊕	○		⊕		⊕	⊕	⊕	⊕
				⊕	+	⊕	⊕	+	⊕	⊕	○	+	○		⊕		⊕	⊕	⊕	⊕
⊕		⊕	⊕	⊕	+	⊕	⊕	⊕	⊕	⊕	⊕	⊕	⊕	⊕	⊕	+	⊕	⊕	⊕	⊕
⊕	○	+[b]	⊕	⊕	+	⊕	⊕	⊕	⊕	⊕	⊕	⊕	⊕	+	⊕	+	⊕	⊕	⊕	⊕
⊕	⊕	+[b]	⊕	⊕	⊕	⊕	⊕	⊕	⊕	⊕	⊕	+	⊕	⊕	⊕	+	⊕	○	⊕	⊕
⊕	⊕	⊕	⊕	⊕	⊕	⊕	⊕	+	⊕	⊕	⊕	○	⊕	+	⊕	⊕	⊕	⊕	⊕	⊕
⊕		+	⊕	+	⊕	⊕	⊕	+	⊕	+	⊕		⊕	+	⊕	+	⊕	⊕	⊕	⊕
⊕		⊕	⊕	+	⊕	○	⊕		⊕	+	⊕		⊕	⊕	⊕		⊕	○	⊕	⊕
○	○	⊕[d]	⊕	⊕	⊕	⊕	⊕	⊕	+	⊕	⊕	⊕	⊕	⊕	⊕	+	○	⊕	⊕	⊕
○	+		⊕	⊕	⊕	○	○	+	⊕	○	⊕	⊕	⊕	⊕	⊕		○	○	⊕	⊕
		+	⊕	○	⊕	○	○	○	⊕	○	⊕		⊕				○	○	⊕	⊕
			⊕	⊕	⊕	○	○	○	⊕	+	○		⊕				○	○	⊕	⊕
⊕	⊕	⊕[d]	○	+	⊕	⊕	⊕	○	⊕	○	⊕		⊕		⊕	⊕	⊕	○	⊕	⊕
⊕	⊕	⊕[d]	○		+	○	○	⊕	○	⊕	○		⊕			+	○	○	⊕	⊕
⊕	⊕	+	⊕	⊕	⊕	⊕	+	+	+		⊕	○	+		⊕	+	+	+	⊕	⊕
○		+	⊕	⊕	+	⊕	+	+	+		⊕		+	+	⊕	+	⊕	+	⊕	⊕
		+	⊕	+	+	⊕	+	+	+		○	+			+	+	⊕	+	⊕	⊕
			+	⊕	+	⊕	+		+		○							+	⊕	⊕
⊕	⊕	+	⊕	⊕	+	⊕	⊕	+	⊕	⊕	⊕	+	⊕	⊕	⊕	⊕	⊕		⊕	⊕
⊕	+	⊕	+	⊕	+	○	○	+	⊕	⊕	○		⊕	⊕	○		⊕		⊕	⊕
○	+	⊕	○	⊕	+	○	⊕	+	+	⊕	○		⊕	⊕	○		⊕		⊕	⊕
⊕	+	⊕	○	⊕	⊕	○	⊕	+	⊕	⊕	○		⊕	⊕	○		⊕		⊕	⊕
+	+	⊕	○	⊕	⊕	○	⊕	+	+	○	⊕	+	+	⊕	⊕	+	⊕	+	⊕	⊕
+	+	+	○	⊕	+	○	○	⊕	+	⊕	⊕		+	⊕	⊕	+	⊕	+	⊕	○

Legend: + - Present in Textbooks
○ - Present in Curriculum Guides

Table 7.1 (continued)
Topics Present in Both Curriculum Guides and Textbooks.

Science Topics	Australia	Austria	Canada	Colombia	Czech Republic	Denmark[e]	Dominican Republic	France	Germany	Greece	Hong Kong	Hungary	Iceland	Ireland
Forces and Motion														
Types of Forces	⊕	⊕	⊕	⊕		○	⊕	○	+	⊕	⊕	⊕	○	⊕
Time, Space, & Motion	○	⊕	⊕	⊕		+	⊕		+	⊕	⊕	⊕	○	⊕
Dynamics of Motion	⊕	○	○	⊕			⊕		+	○		⊕		○
Relativity Theory		○	○	⊕			⊕		+				○	
Fluid Behavior	+	○	+	⊕			⊕			+		○		
Nature or Conceptions of Technology	⊕	○	⊕		⊕	⊕	+	⊕	○			+		○
Interactions of Science, Mathematics, & Technology														
Influence of Math, Technology in Science	⊕	+	⊕		+	○		⊕	○				○	
Applications of Science in Math, Technology	⊕	⊕	⊕		+	○		⊕	⊕	⊕	⊕	⊕	+	⊕
Interactions of Science, Mathematics, & Society														
Influence of Science, Technology on Society	⊕	⊕	⊕	⊕	+	○		⊕	○	+	+	⊕	⊕	⊕
Influence of Society on Science, Technology	⊕		⊕			○		○	○			⊕	+	⊕
History of Science & Technology	⊕	+	⊕	○	+	○		○	+			⊕	+	⊕
Environmental and Resource Issues														
Pollution	⊕	⊕	⊕	+	⊕	○		⊕	○	+		⊕	⊕	⊕
Land, Water, & Sea Resource Conservation	○	⊕	⊕	⊕	⊕	○		⊕	⊕	+	⊕	⊕	⊕	⊕
Material & Energy Resource Conservation	⊕	⊕	⊕	+	⊕	○		⊕	⊕		+	⊕	⊕	○
World Human Population	⊕	⊕	⊕	⊕	⊕	○		○	⊕	+	+	⊕	⊕	⊕
Food Production, Strorage	○	⊕	⊕	+	⊕	○		⊕	○		+	⊕	⊕	⊕
Effects of Natural Disasters	○	○	⊕	⊕	+	○		⊕	○		+	⊕	⊕	⊕
Nature of Science														
Nature of Scientific Knowledge, methods	⊕	○	⊕		+	○		+	⊕		○	○	⊕	⊕
The Scientific Enterprise	○		⊕			○		+	+				⊕	
Science & Other Disciplines														
Science & Mathematics	○	⊕	+			○	○	+	○			○		○
Science & Other Disciplines	○	⊕	+		+		○	○	⊕	○	+	⊕	○	○

a Refer to footnotes to figure 16.4
b The country has requested that this topic be noted as absent in curriculum guides despite the inconsistency with the original coded data.
c The country has requested that this topic be noted as absent in textbooks despite the inconsistency with the original coded data.
d The country has requested that this topic be noted as present in curriculum guides despite the inconsistency with the original coded data.
e New Zealand data include old and new curricula. Caution must be exercised in interpretation since the new curriculum spans all three focal populations, whereas the old curriculum states by-grade objectives that have been separated for each population.
f In Scotland, textbooks are used infrequently in science education. Caution is suggested when comparing contents of guides to the of textbooks.
g The Swedish curriculum guide for science states objectives for all grades from 1 through 9 inclusive, without detailing by-grade objectives. Thus, in this table, curriculum guide data are objectives for grades 1 through 9, while textbook data are only for grade 7 (the upper population of Population 2).
Missing Data: Argentina (no units of appropriate type), Belgium, Bulgaria, People's Republic of China, Iran (no official guides exist), Latvia, Lithuania, Thailand, and Tunisia.

Israel	Japan	Korea	Latvia	Mexico	Netherlands	New Zealand*	Norway	Philippines	Portugal	Romania	Russian Federation	Scotland[i]	Singapore	Slovak Republic	Slovenia	South Africa	Spain	Sweden[g]	Switzerland	USA
⊕		+	⊕	⊕	⊕	⊕		+	⊕		⊕	⊕			⊕	+	⊕	+	⊕	⊕
		+	⊕	⊕	⊕	⊕	⊕	+	+		⊕	⊕			⊕		⊕	⊕	⊕	⊕
		+			⊕	O		+	⊕		O	⊕						⊕	⊕	O
		+	+	⊕	+	O	+				O	⊕			+		+	+	⊕	⊕
			O	⊕	⊕	O	O	+	⊕	⊕	O		⊕	O	⊕		⊕	⊕	+	⊕
⊕	O	⊕	+	⊕	⊕	⊕	+	⊕	+	⊕	⊕	+	O	⊕	⊕	O	+	+	⊕	⊕
	O	⊕	⊕	⊕	⊕	⊕	+	⊕	⊕	+	⊕	+		⊕	⊕		⊕	+	⊕	⊕
	O	O	O	⊕	⊕	O	+	⊕	+	⊕	+	O	⊕	⊕	⊕	O	⊕		⊕	⊕
	O	⊕	O	⊕	⊕	O	+	⊕	+	O	O	O	⊕	O	⊕		⊕	O	⊕	⊕
+		⊕	⊕	⊕	+	⊕	+	⊕	+	⊕	+		+	⊕	⊕		⊕	⊕	⊕	⊕
	O	⊕	⊕	⊕	⊕	⊕	+	⊕	+	⊕	⊕	⊕	⊕	⊕	⊕	⊕	⊕	⊕	⊕	⊕
	O	O	⊕	O	⊕	⊕	+	⊕	+	⊕	⊕	O	⊕	⊕	O	O	⊕	O	⊕	⊕
		⊕	⊕	⊕	⊕	⊕	+	⊕	+	⊕	⊕	O	⊕	⊕	⊕	O	⊕	O	⊕	⊕
		O	⊕	O	⊕	⊕	+	⊕	⊕	⊕	⊕	O	⊕	⊕	⊕	O	O	O	⊕	⊕
		O	⊕	⊕	⊕	⊕	+	⊕	⊕	⊕	⊕	O	⊕	⊕	⊕	O	O	O	⊕	⊕
		⊕	O	O	O	O	+	⊕	+	⊕	O	O	⊕	O	⊕	+	O	O	⊕	⊕
O		⊕	O	⊕	⊕	⊕		⊕		O	O		⊕	⊕	⊕	⊕	⊕		⊕	⊕
		⊕	+	O	⊕	⊕		⊕	O		O		⊕	O	⊕		⊕		+	⊕
O		+	⊕	+	⊕	⊕		⊕	+	⊕		O	O	O			⊕	+	⊕	⊕
		+	⊕	O	⊕	⊕		⊕	+	O	+	O	⊕	⊕	O		⊕	⊕	⊕	⊕

Legend: + - Present in Textbooks
O - Present in Curriculum Guides

for the country. Both symbols can appear independently. If both are present, there is agreement between curriculum guides and textbooks; this agreement probably represents positive evidence that particular curricular attention was intended for that topic in that grade. Circles alone at least suggest that attention was intended to be paid to the topic.

As the table shows, many topics were covered with moderate to high levels of intensity — in the sense of guide and text agreement — in many countries. For instance, 'organs and tissues' was widely covered, as were both plant and animal types, 'sensing and responding,' 'reproduction,' and 'interdependence of life.' In the physical sciences, several topics in 'matter' were covered at moderate to high levels of intensity, as were some topics in 'energy and physical processes.' This was also true of 'applications of science in mathematics, technology,' 'pollution,' and topics in 'conservation of land, water, and sea resources' and 'conservation of material and energy resources.' Canada, Colombia, and the United States showed agreement between guides and textbooks for most topics that they covered at all. Scotland and Russia also showed considerable agreement, although many other topics were present only in curriculum guides. Still other countries (for example, Hong Kong) showed at best modest agreement and only for a limited number of topics.

Focused Versus Diverse Topic Coverage

Instructional time — that is, time in which specific educational opportunities are made available — is, of necessity, limited within any school year. This time has to be distributed among competing topic demands by the specific curricular visions and aims guiding science education. It is likely that a curriculum with many topics devotes less time on average to each — and markedly less to some topics if others receive more of the limited supply of instructional time. Additionally, multiple topics imply shifts in attention among topics; this imposes a further limit on intensity of instruction. On the other hand, a curriculum of few topics per grade level — a *focused curriculum* — would entail fewer competing demands for curricular attention. In such a curriculum, it is likely that all topics would receive more, even if not equal, emphasis.

Recall that figure 5.6 in chapter 5 shows that some countries consistently had more highly focused science curricula than others in the sense that a fewer number of topics were intended for a grade. Even countries with comparatively diverse curricula typically had at least some grade levels in which fewer topics were considered, thereby allowing comparatively more topic emphasis in those years.

Variation in Emphasis in Tracing of Topic Coverage

As noted above, curricular emphasis can be determined based on the number of topics competing for a fixed amount of instructional time within a country. We can make a more sophisticated analysis of curricular emphasis using topic coverage data provided by country experts. TIMSS gathered data from experts in each country on coverage of each framework topic. These experts indicated for each framework topic and each grade level whether the topic was covered, not covered, or was a focus of curricular attention.[1] These responses were then weighted; topics not covered in a grade were assigned a zero weight for curricular attention, those covered and were not a focus received a weight of one, and those that were a focus received a weight of two. All the weights for each grade level's topics within a country were then added together. This total provided an indicator of aggregate demand for curricular attention among the "competing" framework topics within that grade for that country.

Total demand was then used to determine the share or proportion of curricular attention devoted to each topic within that grade and country. This was calculated by dividing zero, one, or two (as appropriate for the particular topic) by the weighted total for that grade and country.[2]

Obviously, a huge amount of data can be generated from this analysis. In the interest of space, attention here is restricted to the upper grade of Population 1. Figure 7.1 presents share or proportion and attention estimates for the upper grade of Population 1. The figure contains data only for a sample of topics particularly relevant to this population. Similarly, in Population 2 there was considerable variation — both among topics and among countries — in the default, "equal share" proportion of aggregate, within-grade curricular attention. This estimate of intensity revealed patterns similar to those of the other approaches.

Variation in Emphasis in Textbook Topics for Populations 1 and 2

Textbook data provide another approach to curricular emphasis. The proportion of a textbook — more precisely, the percentage of blocks into which the book has been partitioned — devoted to a specific topic is one indicator of the emphasis on that topic within a particular grade.

This measure is especially meaningful for those countries that make consistent, extensive use of textbooks in their instructional and learning activities. Note, though, that the simple presence of material in a textbook does not definitely imply its use in a country's classrooms, nor does it take into account any variations in that use. Textbooks can be — and in some countries, often are — supplemented by additional materials from the teacher. Also some material included in the textbook may not be covered. However, the likelihood of coverage is higher when the material is present and less high when it is not. Thus, textbook data are useful in suggesting possible rough bounds on emphasis. This is especially true when attention is restricted to those topics that were treated in both the curriculum guides and textbooks of many countries, and that are thus of demonstrated importance to science education within a particular grade.

This discussion focuses on the topics identified in chapter 6 as widely intended and both supported and emphasized by the corresponding textbooks of the upper grades of Populations 1 and 2. For the upper grade of Population 1 these were 'plants, fungi' and 'animals.' For the upper grade of Population 2, 'organs, tissues' was the only topic that met the criteria.

Figure 7.2 presents data on textbook space for the two commonly intended and emphasized topics in the upper grade of Population 1. The cumulative proportion of textbook space in each country devoted to the topics clearly varied considerably among the countries. A large portion

Figure 7.1
Curricular Attention — Population 1.

The share of curricular attention for each topic within each grade and country was computed. The index of comparative attention is presented by one of the five symbols. They are for the upper grade of Population 1 for some widely intended topics in that grade and for all countries. This illustrates the variation typical of the data for all topics.

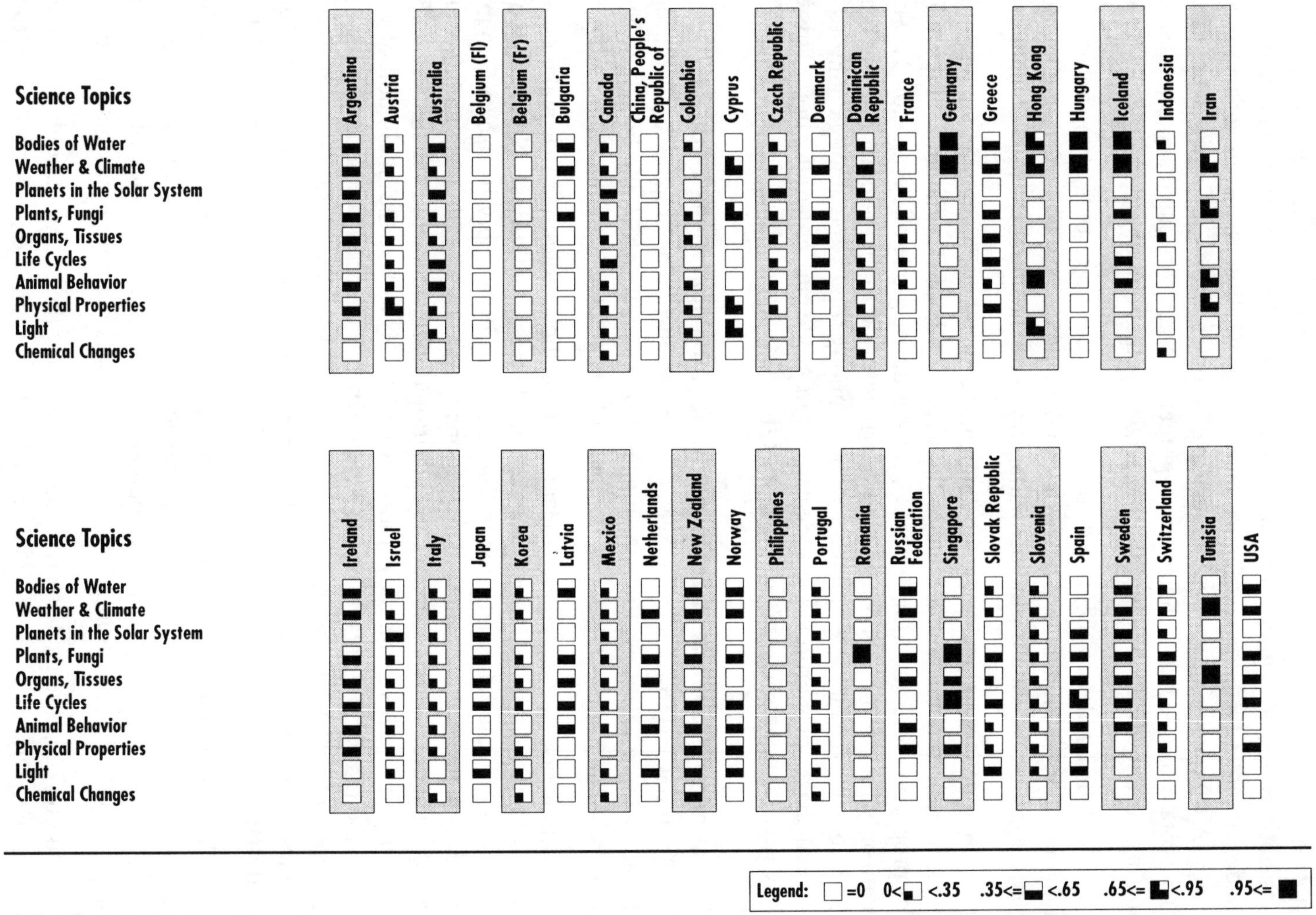

Science Topics

Bodies of Water
Weather & Climate
Planets in the Solar System
Plants, Fungi
Organs, Tissues
Life Cycles
Animal Behavior
Physical Properties
Light
Chemical Changes

Science Topics

Bodies of Water
Weather & Climate
Planets in the Solar System
Plants, Fungi
Organs, Tissues
Life Cycles
Animal Behavior
Physical Properties
Light
Chemical Changes

Legend: □ =0   ▫ 0<■ <.35   ▣ .35<= ■ <.65   ◼ .65<= ■ <.95   ■ .95<=

Figure 7.2
Variations in Textbook Space for Two Commonly Intended Topics - Population 1.
The percentage of textbook blocks devoted to the two emphasized and widely covered topics at the upper grade for Population 1 is shown for each country. The variation in attention to these topics is obvious.

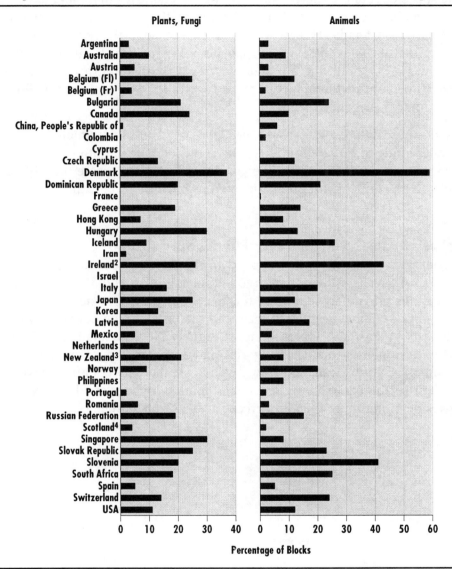

Percentage of Blocks

1   The National Research Coordinators of Belgium have collected data only from curriculum guides. Due to the great level of detail of the guides and their extensive use, data from these are compared in this display with the textbook data supplied from all other countries.
2   Ireland: Data applies to lower grade of Population 1.
3   New Zealand: Since textbooks are not used in science instruction at this level, data are collected from teacher resource booklets. It is important to bear in mind that teachers use their own discretion in choosing units and topics to teach at this level.
4   Although Scottish researchers made every effort to include data on a variety of textbooks that were deemed representative of textbooks in their country, their infrequent use in science education suggests caution when interpreting results.
Missing Data: Germany, Lithuania, Sweden, and Tunisia.

of this material in virtually every country was devoted to 'plants, fungi' and 'animals,' although Denmark, Ireland, and Slovenia stand out as having devoted a particularly large proportion of space to these topics. Austria, Colombia, Portugal, Scotland, and French-speaking Belgium had comparatively little space devoted to these topics among the countries that included them. The relative proportion devoted to each topic also varied among the countries. Countries such as Flemish-speaking Belgium, Canada, Hungary, and Singapore devoted more space to 'plants, fungi' relative to 'animals,' — whereas the opposite was true for Denmark, Ireland, the Netherlands, and Slovenia. Overall, there is much diversity, with countries such as Denmark devoting more than 70 percent of the books to these two topics, whereas Iran, Colombia and Portugal devoted less than 5 percent to them.

Figure 7.3 presents these data for 'organs, tissues,' the only commonly intended and emphasized topic in the upper grade of Population 2. Clearly, considerable variation among the countries in the emphasis was accorded this topic. Seven countries devoted more than 10 percent of their textbook blocks to this topic; most others devoted fewer than 10 percent of their textbook blocks to it.

Variation in Emphasis in Textbook Topics for Physics Specialists

The textbook content analysis was also applied to the physics specialist texts of TIMSS Population 3. Figure 7.4 presents the cumulative proportion of textbooks devoted to the six commonly intended topics previously identified for this population. Considerable intercountry variations in the amount of emphasis given to these topics exists. Some countries were unique in the level of emphasis given these topics — most notably New Zealand, Israel, and South Africa. Others (for example, Romania, Bulgaria, and Denmark) put comparatively little emphasis on these topics. Cyprus, Iran, and New Zealand are the only countries devoting more than 25 percent of their textbooks to 'time, space, and motion,' while Israel and New Zealand are the only countries devoting more than 25 percent to 'wave phenomena.'

Figure 7.3

Variations in Textbook Space for 'Organs and Tissues' - Population 2.

The life science topic of 'organs and tissues' is the single topic widely covered and emphasized in textbooks at the upper grade of Population 2. Here the percentage of textbook blocks devoted to this topic is shown for all countries providing data. Five countries devoted more than fifteen percent of textbook blocks to this topic; eight devoted fewer than five percent.

Percentage of Blocks

[1]   Refer to footnotes in figure 7.2.

Note: The following countries submitted data for Population 2 textbooks but did not cover this topic: Czech Republic, Denmark, France, Germany, Iran, and Lithuania.

Figure 7.4
Variations in Textbook Space for Six Core Topics — Physics Specialists.

Six topics — five of them in the area of energy and physical processes, one in the area of forces and motion — were widely covered and emphasized for the Physics Specialists of Population 3. The textbook block percentages for these four topics presented here show the extensive variation even in comparatively focused textbooks.

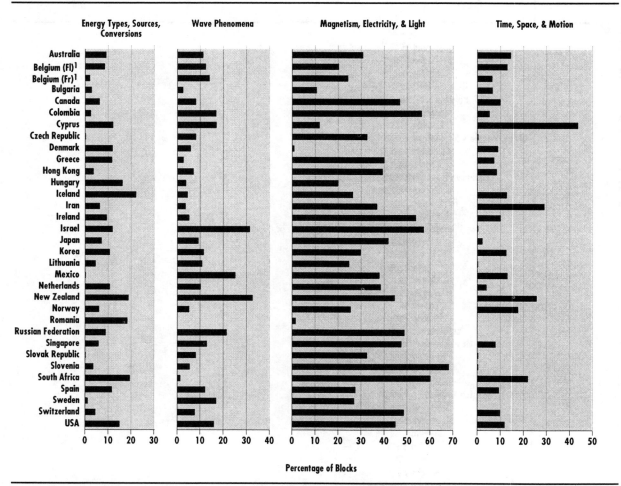

1   The National Research Coordinators of Belgium have collected data only from curriculum guides. Due to the great level of detail of the guides and their extensive use, data from these are compared in this display with the textbook data supplied from all other countries.

Missing Data: Argentina, Austria, People's Republic of China, Dominican Republic, France, Germany, Italy, Latvia, Philippines, Portugal, Scotland, and Tunisia.

NOTES

[1] "Focus" and "coverage" were not precisely defined, but they provide at least rough indications of curricular attention levels.

[2] Using this method, some topics within a grade and country could receive disproportionate shares of curricular attention; over all topics, however, this default "equal share" proportion of attention estimates the true share across many topics. Note that, to avoid decimal fractions, the indices were multiplied by 10.

Chapter 8

OTHER ASPECTS OF THE CURRICULUM: PERFORMANCE EXPECTATIONS AND DISCIPLINARY PERSPECTIVES

Students are expected not only to gain the scientific knowledge discussed in the previous chapters, but also to think scientifically and do things with science content. These *student performance expectations* are central to countries' curricular visions and intentions. They are a second, fruitful source for identifying shared, and different, curricular aims. Data on student performance expectations — on what students are expected to be able to do with science at particular points in their schooling — were gathered from analysis of country curriculum guides and textbooks. These expectations are set forth in the second aspect of the Third International Mathematics and Science Study (TIMSS) science framework (see appendix C).

Most of this chapter is devoted to exploring these student performance expectations. In addition, a brief section is included on the third TIMSS science framework aspect, *disciplinary perspectives*. This aspect covers broad perspectives on, perceptions of, and attitudes toward science — that is, what students are supposed to think or feel about science as a discipline. These data also were obtained from curriculum guides and textbooks; however, because many countries did not address this aspect in their curricular materials, the analysis is limited.

Commonly Intended Performance Expectations

To determine common performance expectations across countries, the units and blocks of the sampled curriculum guides and textbooks were coded according to the categories of the TIMSS science framework performance expectations. The most commonly intended performance expectations for the upper grades of Populations 1 and 2 are shown in table 8.1. The table lists the student performance expectations that were found in the curriculum guides for the indicated grade levels in at least 70 percent of the TIMSS countries. Topics found *only* in curriculum guides — without mention and support in the science textbooks for the grade level in question — are in the top pair of lists. Topics found in curriculum guides *and* supported by inclusion in the appropriate textbooks are in the middle pair of lists. Performance expectations included widely in textbooks (70 percent) but not included widely in the corresponding curriculum guides are in the final list.

The range of expectations for the upper grade of Population 1 was quite reduced and included some aspects of 'understanding,' 'using tools, routine procedures, and science processes,' and 'investigating the natural world' — although the latter expectations were restricted to mention in curriculum guides without support in 70 percent of textbooks. If attention is restricted only to those expectations in curriculum guides *and* supported by textbooks, the common expectations were even more limited, excluding *all* aspects of 'investigating the natural world.' Perhaps even more startling is that the only emphasized expectation was 'understanding simple information.'[1] What appears to have been widely held intentions of developing a balanced array of expected performance abilities became more limited when textbook support was examined — rather than accepting the official intentions presented in curriculum guides alone — and became severely limited to 'understanding simple information' and when textbook emphasis was considered.

These data point to a clash of visions between the curriculum guides and textbooks of many countries. Although the curriculum guides cite some demanding expectations for Population 1 students, the visions held by textbook makers were considerably less demanding as inferred from their products and were limited mostly to procedural knowledge. This is exactly the same situation that was encountered in analyzing curriculum guides and textbooks in mathematics.

This clash of visions was somewhat less for students in the upper grade of Population 2. The performance expectations of 'conducting investigations' and 'interpreting data'[2] that were in curriculum guides for the upper grade of Population 1 are now supported in textbooks. The performance expectations supported in textbooks in the upper grade of Population 1 continue to be supported in textbooks intended for the upper grade of Population 2. There are a number of performance expectations from the general categories of 'theorizing, analyzing, and solving problems,' 'using tools, routine procedures, and science processes' (all specific categories), and 'communicating' that are called for by guides and supported by textbooks. On the other hand, only two performance expectations under 'investigating the natural world' are supported by textbooks: 'conducting investigations' and 'interpreting data.' The other three performance expectations under this general category, 'identifying questions to investigate,' 'designing investigations,' and 'formulating conclusions from data,' are mentioned in curriculum guides but they are not largely supported by textbooks, which seems to indicate again a more conservative set of expectations for student performances in textbooks than in curriculum guides.

This conservative nature of textbooks is further evident when examining the two performance expectations that are *emphasized* in textbooks. These are 'understanding simple information' and 'understanding complex information.'[3] Thus, even in a context of rich and varied demands presented in curriculum guides and supported by texts, the common emphases in textbooks remained on more routine aspects of 'understanding.' The only progression in this regard between Population 1 and Population 2 is between understanding 'simple' information and understanding 'complex' information.

Table 8.1

Commonly Intended Performance Expectations.

Performance expectations indicated in the curriculum guides of 70 percent of TIMSS countries, 70 percent of countries' textbooks, or both are shown here. Those emphasized in textbooks (accounting for 6 percent or more of textbook blocks) are marked with an asterisk. These data are only for upper grades of Populations 1 and 2.

POPULATION 1	POPULATION 2
Curriculum Guides (Not in Textbooks):	
Investigating the Natural World Conducting Investigations Interpreting Data	**Theorizing, Analyzing, & Solving Problems** Making Decisions **Investigating the Natural World** Identifying Questions to Investigate Designing Investigations Formulating Conclusions From Data **Communicating** Sharing Information
Curriculum Guides (Included in Textbooks):	
Understanding Understanding Simple Information* Understanding Complex Information **Using Tools, Routine Procedures, and Science Processes** Using Apparatus, Equipment, Computers Doing Routine Experimental Operations Gathering Data	**Understanding** *Understanding Simple Information *Understanding Complex Information Understanding Thematic Information **Theorizing, Analyzing, & Solving Problems** Abstracting & Deducing Scientific Principles Applying Scientific Principles to Solve Quantitative Problems Applying Scientific Principles to Develop Explanations Constructing, Interpreting, & Applying Models **Using Tools, Routine Procedures, & Science Processes** Using Apparatus, Equipment, Computers Doing Routine Experimental Operations Gathering Data Organizing & Representing Data Interpreting Data **Investigating the Natural World** Conducting Investigations Interpreting Data **Communicating** Accessing & Processing Information
Exclusively in Textbooks:	
Using Tools, Routine Procedures, & Science Processes Organizing & Representing Data	

*Perfomance expectation emphasized in textbooks.

It is worth noting that the analysis of *mathematics* textbooks also uncovered an essentially conservative approach to performance expectations in comparison to that present in curriculum guides.

Variations in Intended Performance Expectations

Even though the TIMSS countries held many student performance expectations in common, there was variation in individual country expectations as reflected in their curriculum guides (see table I8.2 in appendix I for data on all countries and table 8.2 in this chapter for representative countries). Unweighted curriculum guide data indicate the presence or absence of specific performance expectations. Some countries — for example, Greece — showed diverse, extensive expectations at the upper grade of Population 1. In others — for example, Japan — the variety of expectations was considerably more restricted.

More of the possible expectations from the framework were used in the upper grade of Population 2. For some countries — Canada and New Zealand, for example — a wide range of expectations was present at both grade levels. For Population 1, Flemish-speaking Belgium included all sub-categories from all performance expectations with the exceptions being 'theorizing, analyzing, and problem solving' and 'communicating.' However, it included the full range of all performance expectation categories for Population 2. Similarly, Ireland included some performance expectations ('using tools, routine procedures, and science processes,' 'investigating the natural world,' and 'communicating') for Population 1 but a fuller variety for Population 2. For a few countries — for example, Greece and Spain — a more diverse set of Population 1 expectations was matched by a less diverse — or, perhaps better, more focused — set of Population 2 expectations.

Variations in Textbook Emphasis of Performance Expectations

Textbook data, although only a secondary reflection of curricular intentions, can be weighted to show relative emphasis. Thus, the percentage of a textbook's blocks that were coded as covering a particular performance expectation indicate the emphasis on that expectation in that textbook. Aggregated to the country level, these data provide an indication of relative emphases within a country and of variations among countries. These data are necessarily more diverse than the "presence versus absence" data from curriculum guides; they must be interpreted remembering that the roles and uses of textbooks differ among countries.

Figure 8.1 presents the percentage of textbook blocks that included each performance expectation for the upper grade of Population 1 for selected countries. Corresponding data for the upper grade of Population 2 are presented in figure 8.2. Overall, although the degree of variation was high, the percentages for many categories were surprisingly low — which is a further reflection of the more conservative, less rich vision of student performance expectations typical of textbooks as seen in table 8.1.

The data for Population 1 present some interesting variations. 'Understanding simple information' represents from less than 20 percent of blocks in the textbooks of New Zealand and Flemish-speaking Belgium, to near 100 percent of blocks in textbooks from Greece. Regarding 'gathering data,' some countries have no blocks characterized by this performance expectation in their textbooks (as in Greece and Latvia), whereas other countries have a sizable amount of such blocks (as in Flemish-speaking Belgium and New Zealand).

Data in this table corroborate the general finding of a more conservative set of performance expectations present in textbooks than in curriculum guides, similar to what was found for mathematics textbooks. Although curriculum guides appear to call for performance expectations of 'theorizing, analyzing, and solving problems' and also 'investigating the natural world,' very little is present in textbooks to support them. There are some notable exceptions: Japan and Flemish-speaking Belgium appear to do quite a bit more in the area of 'identifying questions to investigate' and New Zealand and Flemish-speaking Belgium in 'conducting investigations.' New Zealand is also noteworthy in the extent to which textbooks contain blocks focusing on the categories under 'communicating.'

For Population 2, the conservative trend in textbooks appears even greater. What little was done in textbooks for the upper grade of Population 1 in 'theorizing, analyzing, and solving problems' and 'investigating the natural world' is largely no longer present in those textbooks intended for the upper grade of Population 2. For example, there are very few blocks on 'identifying questions to investigate' in Japanese textbooks at this level, where these blocks had been more than 30 percent in the textbooks intended for the upper grade of Population 1. New Zealand textbooks at this level have about one-half of the percentage of blocks (17 percent) devoted to 'conducting investigations' than had been the case for the upper grade of Population 1 (34 percent). This trend is particularly striking as the scientific content at this level is more complex and curriculum guides include a wider variety of performance expectations.

It may be the case that this pattern is the result of the increased complexity of the subject matter, making textbooks concentrate on the presentation of facts and explanations. In such a case, 'investigating the natural world' might diminish in importance. All required laboratory manuals were included in these analyses. However, it may be precisely in the area of investigation where teachers are not relying on standard sets of materials that are the focus of this report.

Variation in the percentage of blocks devoted to 'understanding simple information' is also quite high in textbooks for the upper grade of Population 2, ranging from 1 to 10 percent (Flemish-speaking Belgium) to more than 80 percent (Greece, Hungary, Ireland, and Norway).

Relationships Between Performance Expectations and Content Topics

To this point, performance expectation data have been considered independently of science topic. While performance expectations have been examined for individual countries and across

Table 8.2
Variations in Performance Expectations.
The presence or absence of each performance expectation is indicated for the upper grades of Populations 1 and 2. These data are based only on curriculum guides. The data show that although common expectations existed across countries, they did so in a context of strong cross-national variations.

	Understanding		
	Simple Information	Complex Information	Thematic Information
Belgium (Fl)	1/2	1/2	1/2
Canada	1/2	1/2	1/2
Czech Republic	1/2	1/2	2
France	1/2	1/2	1/2
Greece	1/2	1	1
Hungary	1/2	1/2	1/2
Ireland	2	2	2
Japan	1/2	2	2
Latvia	1/2	2	1/2
New Zealand*	1/2	1/2	1/2
Norway	1/2	1/2	-
Spain	1/2	1	1

	Using Tools, Routine Procedures, & Science Processes				
	Using Apparatus, Equipment & Computers	Conducting Routine Experimental Operations	Gathering Data	Organizing & Representing Data	Interpreting Data
Belgium (Fl)	1/2	1/2	1/2	1/2	1/2
Canada	1/2	1/2	1/2	1/2	1/2
Czech Republic	2	2	1/2	-	-
France	1/2	1/2	1/2	1/2	1/2
Greece	1	1/2	1	1	1
Hungary	1/2	1/2	1/2	2	1/2
Ireland	1/2	1/2	1/2	1/2	1/2
Japan	2	2	2	2	2
Latvia	2	1/2	1/2	2	1/2
New Zealand*	1/2	1/2	1/2	1/2	1/2
Norway	1/2	1/2	1/2	1/2	1/2
Spain	1/2	1/2	1/2	1/2	1/2

* New Zealand: TIMSS analyses were based on data from the 1989 revised edition of the curriculum guides that were valid until 1995. New guides were published in 1993 and schools were advised to work toward their implementation at the beginning of 1995.

Theorizing, Analyzing, & Problems

	Abstracting & Deducing Scientific Principles	Applying Scientific Principles... to Solve Quantitative Problems	Applying Scientific Principles... to Develop Explanations	Constructing, Interpreting, & Applying Models	Making Decisions
Belgium (Fl)	2	2	2	2	2
Canada	1/2	1/2	1/2	1/2	1/2
Czech Republic	1	2	1/2	-	-
France	2	1/2	2	1/2	2
Greece	1	1/2	1	1	1
Hungary	1/2	2	2	2	2
Ireland	2	2	2	2	2
Japan	2	2	2	2	2
Latvia	2	2	2	2	2
New Zealand*	1/2	1/2	1/2	1/2	1/2
Norway	1/2	-	-	2	1/2
Spain	-	-	-	1/2	-

Investigating the Natural World / Communicating

	Identifying Questions to Investigate	Designing Investigations	Conducting Investigations	Interpreting Investigational Data	Formulating Conclusions from Investigational Data	Accessing & Processing Information	Sharing Information
Belgium (Fl)	1/2	1/2	1/2	1/2	1/2	2	2
Canada	1/2	1/2	1/2	1/2	1/2	1/2	1/2
Czech Republic	-	-	1/2	-	-	2	2
France	1/2	1/2	1/2	1/2	1/2	2	2
Greece	1	1	1	1	1	1	1
Hungary	1/2	2	2	1/2	2	1/2	2
Ireland	1/2	1/2	1/2	1/2	1/2	1/2	1/2
Japan	1/2	2	2	2	2	2	-
Latvia	2	1/2	1/2	1/2	1/2	2	-
New Zealand*	1/2	1/2	1/2	1/2	1/2	1/2	1/2
Norway	1/2	1/2	1/2	1/2	1/2	1/2	1/2
Spain	-	1	1	1	-	1/2	1/2

Legend:	1	Present in the curriculum guide for the upper grade of Population 1
	2	Present in the curriculum guide for the upper grade of Population 2
	-	Not present in either

Figure 8.1
Variation of Textbook Use of Performance Expectations - Population 1.

The textbook block percentages for various performance expectations are shown for the upper grade of Population 1. The data are for all expectations for a sample of countries. The data again show considerable variation.

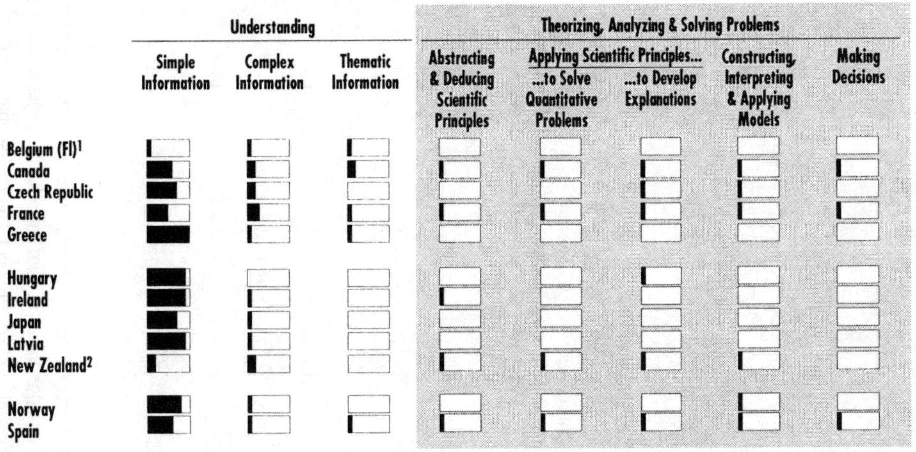

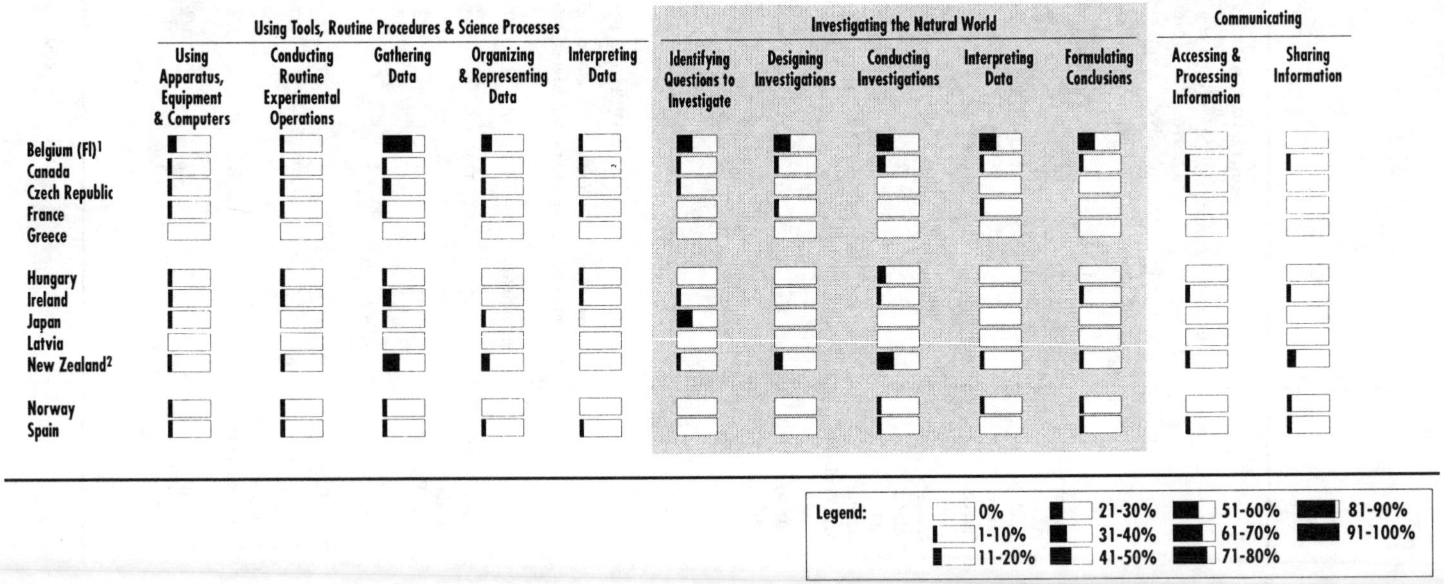

1 The National Research Coordinators of Belgium have collected data only from curriculum guides. Due to the great level of the detail of the guides, and their extensive use, data from these are compared in this display with the textbook data supplied from all other countries.

2 New Zealand: Since textbooks are not used in science instruction at this level, data are collected from teacher resource booklets. It is important to bear in mind that teachers use their own discretion in choosing units and topics to teach at this level.

Figure 8.2
Variation of Textbook Use of Performance Expectations - Population 2.

The textbook block percentages for various performance expectations are shown for the upper grade of Population 2. Similar variations as with Population 1 persisted at this level.

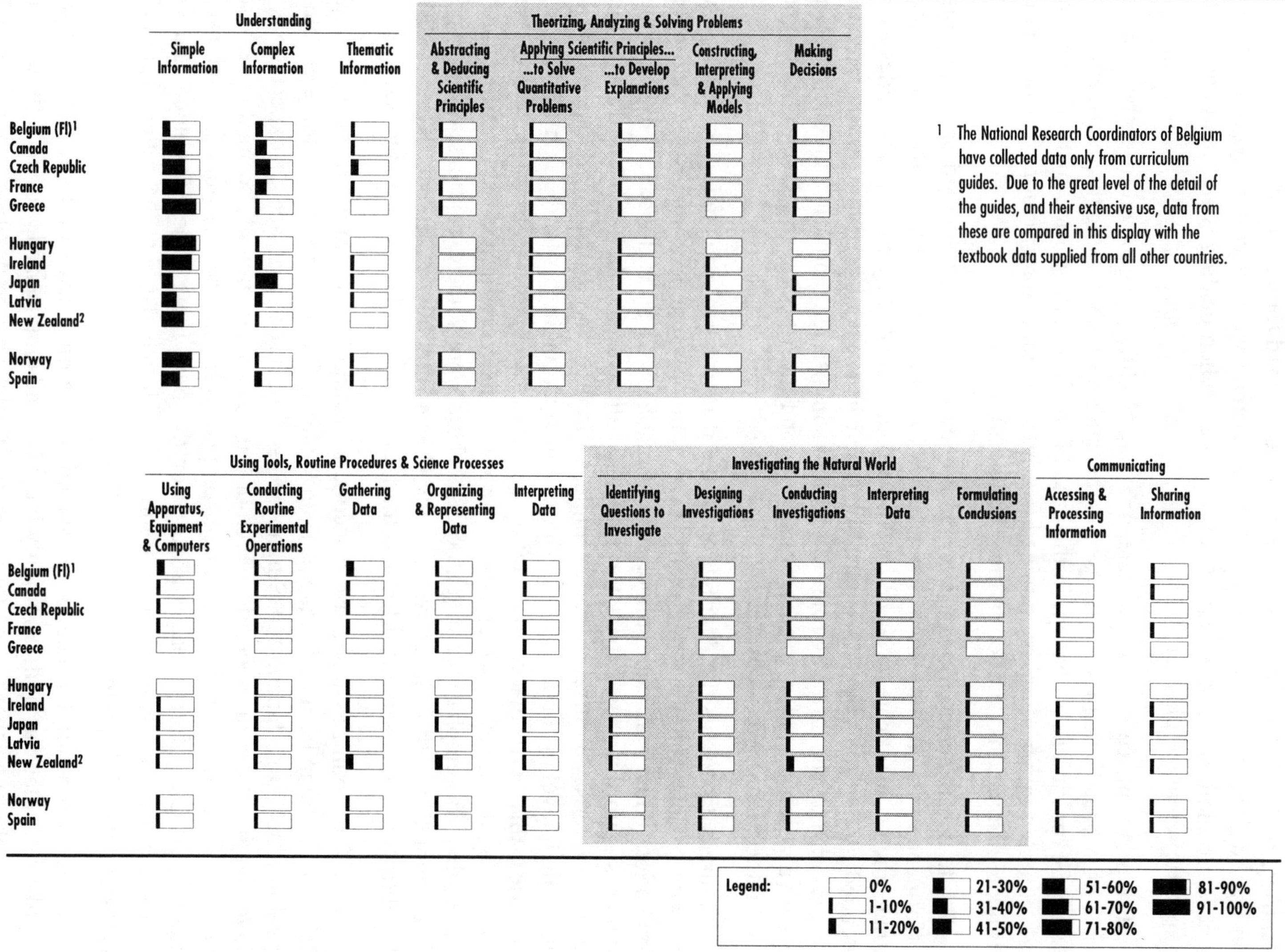

1 The National Research Coordinators of Belgium have collected data only from curriculum guides. Due to the great level of the detail of the guides, and their extensive use, data from these are compared in this display with the textbook data supplied from all other countries.

countries, performance expectations are always examined for all science topics pooled. This section uses weighted textbook block data to determine whether certain types of student performances are more frequently expected than others when dealing with broad content categories or specific science topics.

The performance expectation categories are slightly reorganized from those presented earlier into six broad categories to be used in the analyses below, using the intermediate framework levels of expectations — except in the case of 'understanding' for which 'simple information' and 'complex information' are retained (so little 'thematic information' was present that it was eliminated from consideration in these analyses).

While it was necessary to use textbook data to compute weighted emphasis estimates for performance expectations, the science topics to which they were related in the analyses were either broad content categories or specific topics indicated by curriculum guides as commonly intended among the countries and emphasized in textbooks. Even so, interpretation should always consider the differing roles of textbooks in different countries, the seemingly more conservative vision of student expectations common for textbooks, and the choice of topics emphasized rather than merely covered.

Figure 8.3 presents the distribution of the performance expectations categories when dealing with 'plants, fungi' — one of the two commonly intended topics for the upper grade of Population 1 that was both supported and emphasized by textbooks. How did the different countries apportion their expectations for what students should be able to do with this content? As shown, most countries — for example, Colombia, the Russian Federation, and Iceland — used a high level of 'understanding simple information' in this textbook material. In almost all countries, this performance expectation accounted for more than 40 percent of the textbook material on this topic — which is in accord with the more limited expectations that seem typical of what was emphasized in textbooks. There were some exceptions to this expectation — for example, in both Belgian systems, the Czech Republic, Singapore, Korea, and the Slovak Republic, the topic accounted for less than 20 percent of the textbook blocks.

Very few countries made use of 'communication' expectations (except, for example, the Dominican Republic and Singapore), and with a few exceptions — for example, Flemish-speaking Belgium, Japan, and Singapore — little use was made of expectations related to 'investigating the natural world.' Much of this material commonly seemed to be based on simple factual knowledge rather than more complex performances. A similarly high use of 'understanding simple information' was seen for the next most widely intended and emphasized Population 1 topic: 'animals' (see figure 8.4). Notice also that the pattern here is essentially the same noted for 'plants, fungi' (see figure 8.3).

In figure 8.5, similar data are presented for the only textbook topic for the upper grade of Population 2 that was both commonly intended and emphasized: 'organs, tissues.' Here it can again be observed that the majority of the blocks in most textbooks are devoted to 'under-

Figure 8.3

Performance Expectations for 'Plants, Fungi.'

The percentages of performance expectations in textbook blocks dealing with 'plants, fungi' are presented for the upper grade of Population 1. In almost all countries, 'understanding simple information' accounted for close to 40 percent or more of textbook blocks.

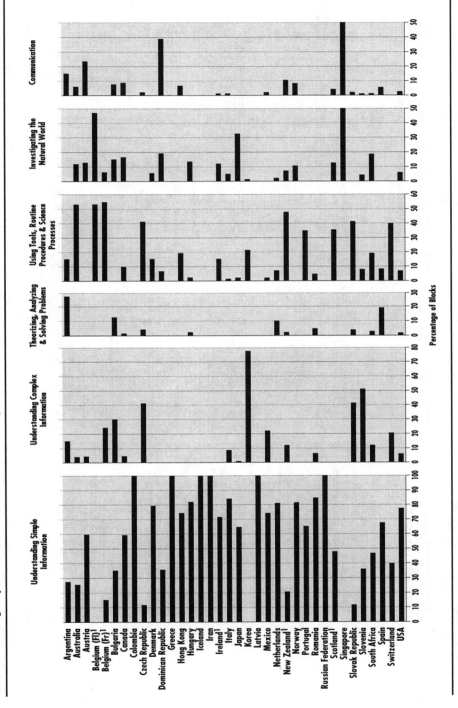

Percentage of Blocks

1   Refer to footnotes in figure 18.1.

Note: Five countries which provided Population 1 textbooks did not cover this topic: China, Cyprus, France, Israel, and Philippines.

Missing Data: Germany, Lithuania, Sweden, and Tunisia.

Figure 8.4

Performance Expectations for 'Animals.'

Data similar to those for the previous figure are presented for the next most common topic for the upper grade of Population 1. Again, considerable emphasis was placed on 'understanding simple information.'

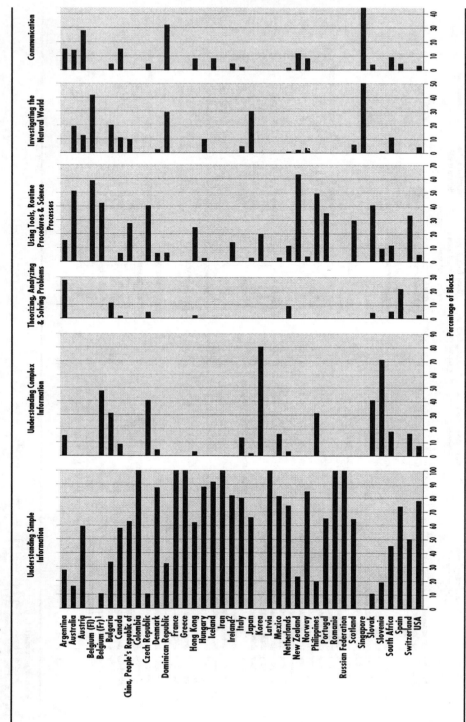

1   Refer to footnotes in figure l8.1.

Note: Two countries which provided Population 1 textbooks did not cover this topic: Cyprus and Israel.

Missing Data: Germany, Lithuania, Sweden, and Tunisia.

Figure 8.5

Performance Expectations for 'Organs, Tissues.'

Data similar to those in Figure 8.3 are presented on the most common topic for the upper grade of Population 2. Again, greater emphasis is shown on 'understanding simple information,' although there is considerable more emphasis on 'understanding complex information' than had been the case with the most commonly covered topics for the upper grade of Population 1.

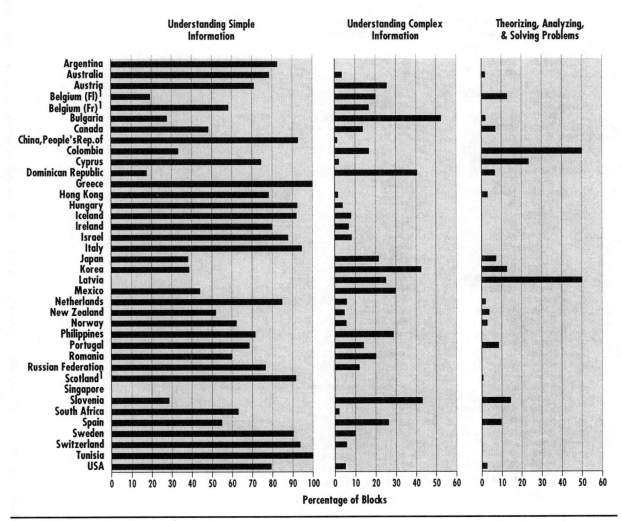

Note: The following countries provided Population 2 textbook data but did not cover this topic: Czech Republic, Denmark, France, Germany, Iran, Lithuania, and Slovak Republic.

Figure 8.5 (continued)
Performance Expectations for 'Organs, Tissues.'

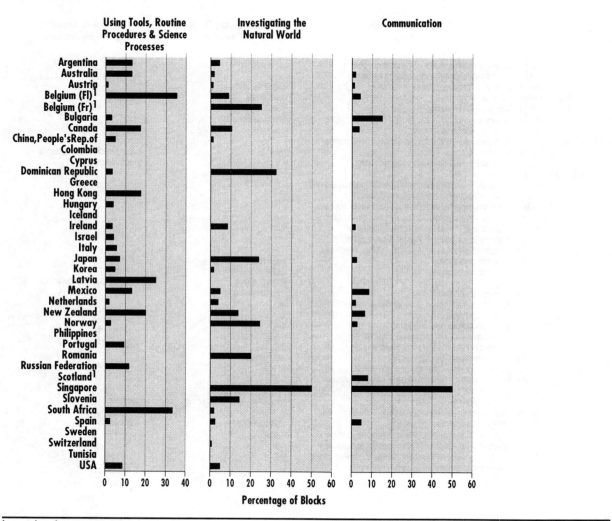

Note: The following countries: Czech Republic, Denmark, France, Germany, Iran, Lithuania, and Slovak Republic, provided Population 2 textbook data but did not cover this topic.

standing simple information' (Greece, Tunisia, and Scotland are good examples) or to a combination of this topic and 'understanding complex information' (such as the case of Austria, Bulgaria, and Korea). Only Colombia and Latvia show a sizable amount of 'theorizing, analyzing, and solving problems.' French-speaking Belgium, the Dominican Republic, Japan, Norway, and Singapore are similar in that a notable amount of 'investigating the natural world' appears to complement the predominant expectation of 'understanding.' Singapore is unique in its combined emphasis on 'investigating the natural world' and 'communication.'

Performance Expectations for Physics Specialists

The analyses conducted for the upper grades of Populations 1 and 2 were also performed for the physics specialists in Population 3. Table 8.3 is the Population 3 physics specialist counterpart of table 8.1.

What was expected of these students was more diverse. Curriculum guides included performance expectations from all categories with the exception of 'communicating.' Although curriculum guides call for investigating the natural world through 'conducting investigations' and 'interpreting data,' these expectations are not commonly supported in textbooks. It is important to note that countries coded required laboratory manuals, and it is therefore unlikely that these expectations lie in other commonly used documents not analyzed. However, it is also more likely that teachers themselves, through optional ancillary materials and other instructional practices, may be implementing performance expectations mentioned in guides yet only marginally supported to a substantial degree in textbooks. What textbooks do emphasize at this level appears to be the understanding of factual information and the solving of paper and pencil exercises. Again, as was the case for the upper grades of Population 1 and 2, textbooks appear to be more conservative in the performance expectations they contain than were curriculum guides.

Figure 8.6 presents textbook proportion — and thus variation — data for a selection of individual countries (see appendix I, figure I8.6 for all countries). In general, many countries emphasized 'understanding' — especially the United States and, to a lesser extent, Australia and Canada. Some notable exceptions were seen as well — for example, the Russian Federation's and Spain's notable emphasis on 'constructing, interpreting, and applying models.' Switzerland was unique in having more than 11 percent of blocks devoted to 'conducting investigations.' Some countries devoted much less space to quantitative problem solving (the Russian Federation, the United States, and Australia) than others (Canada, Greece, Japan, and Spain). Textbooks remained less varied and less demanding than did curriculum guides.

Perspectives

Many textbooks and curriculum guides contained broad-based statements about the nature of scientific habits of mind, the everyday uses of science (presumably included to increase student interest), careers in fields that made use of the sciences, and attitudes towards science and its importance. In TIMSS, these perspectives were formed into a set of six categories. Sampled curriculum guides and textbooks were coded for their direct use of these perspectives.

Table 8.3

Commonly Intended Performance Expectations — Physics Specialists.

The performance expectations indicated in the curriculum guides of 70 percent of TIMSS countries, 70 percent of countries' textbooks, or both are shown here. Expectations emphasized in textbooks (that is, accounting for 6 percent or more of textbook blocks) are marked with an asterisk. These data are only for materials related to the Physics specialists of Population 3.

POPULATION 3

Curriculum Guides (Not in Textbooks):

Using Tools, Routine Procedures, and Science Processes
    Using Apparatus, Equipment, Computers
    Doing Routine Experimental Operations
    Gathering Data
    Organizing and Representing Data
    Interpreting Data

Investigating the Natural World
    Conducting Investigations

Curriculum Guides (Included in Textbooks):

Understanding
    Understanding Simple Information*
    Understanding Complex Information*

Theorizing, Analyzing, and Solving Problems
    Abstracting and Deducing Scientific Principles
    Applying Scientific Principles to Solve Quantitative Problems*
    Applying Scientific Principles to Develop Explanations

Exclusively in Textbooks:

Theorizing, Analyzing, and Solving Problems
    Constructing, Interpreting, and Applying Models

*Perfomance expectation emphasized in textbooks.

Figure 8.6

Variations in Textbook Use of Performance Expectations — Physics Specialists.

The textbook block percentages for various performance expectations are shown for books used by the physics specialists of Population 3. The data are shown for all performance expectations but for a representative selection of countries. A general emphasis on understanding simple and complex information was seen in many countries as well as considerable emphasis on solving quantitative problems.

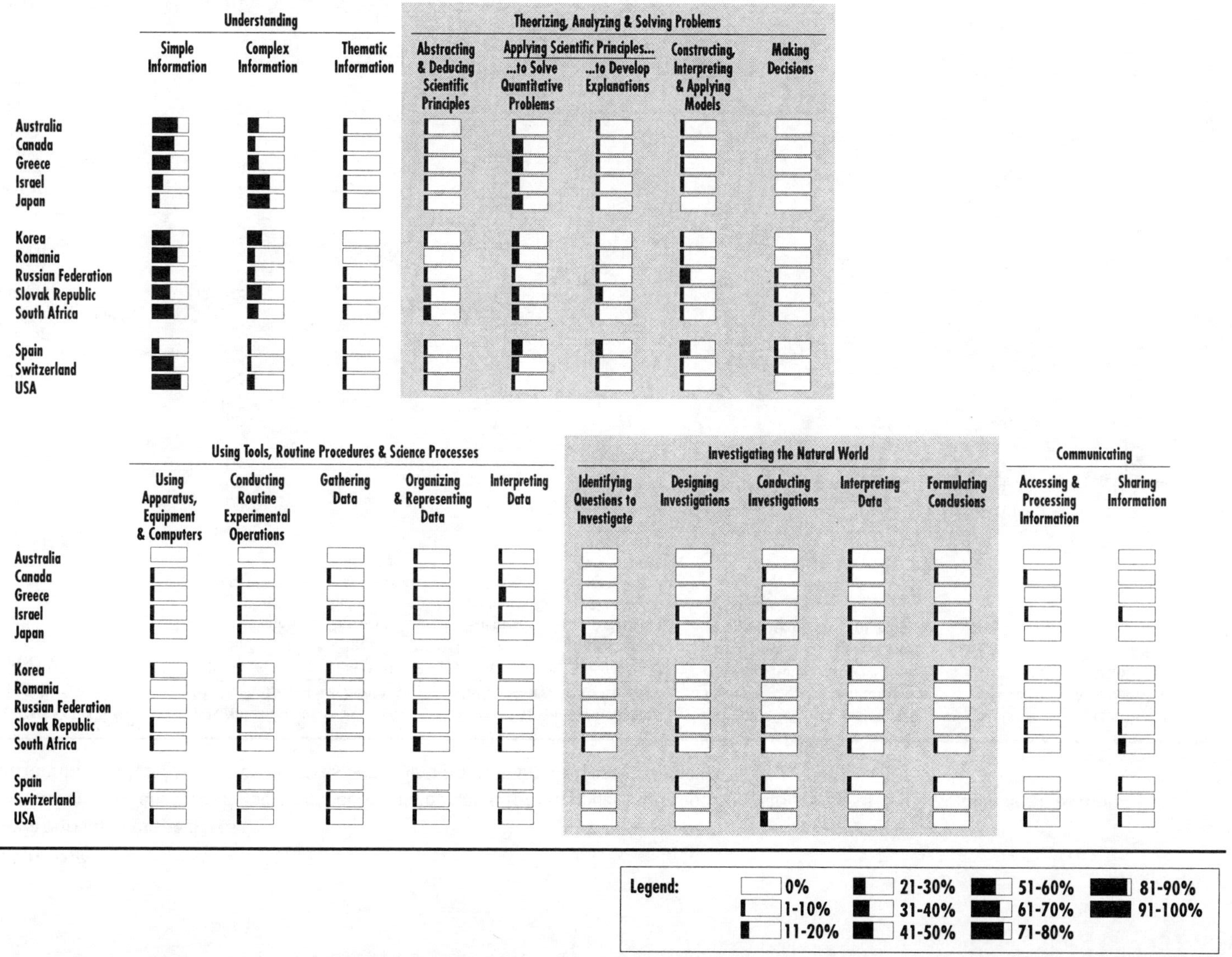

Table 8.4
Variations in Perspectives.

The presence or absence of each perspective is indicated for the upper grades of Populations 1 and 2. These data are based on curriculum guides. Use of perspectives was common in many of these documents.

Perspectives of Populations 1 and 2 — Country	Attitudes Towards Science, Mathematics & Technology		Careers Involving Science, Mathematics, & Technology		Participation in Science & Mathematics by Under-represented Groups	Science, Mathematics & Technology to Increase Interest	Safety in Science Performance	Scientific & Mathematical Habits of Mind
	Positive Attitudes Towards Science, Mathematics & Technology	Skeptical Attitudes Towards Science, Mathematics & Technology	Promoting Careers in Science, Mathematics & Technology	Promoting the Importance of Science, Mathematics & Technology in Non-technical Careers				
Australia	1/2	1/2	1/2	1/2	1/2	1/2	1/2	1/2
Austria	1/2	2	2	2	1/2	1/2	1/2	1/2
Belgium (Fl)	1/2	-	-	-	-	2	2	1/2
Belgium (Fr)	-	1/2	-	-	-	1/2	-	1/2
Canada	1/2	-	2	1/2	1/2	1/2	1/2	1/2
Colombia	1/2	-	2	2	-	1/2	1/2	1/2
Czech Republic	-	-	2	2	-	-	-	2
Denmark[1]	-	-	2	2	-	-	2	-
Dominican Republic	-	-	-	-	-	-	-	-
France	1/2	-	2	2	-	1/2	2	1/2
Germany[1]	-	-	-	-	-	-	-	-
Greece	1/2	-	-	-	-	1/2	1/2	1/2
Hong Kong	2	2	2	2	-	1/2	2	-
Hungary	1/2	1/2	2	2	-	2	2	-
Iceland	1/2	1/2	-	-	-	1/2	1/2	1/2
Ireland	2	-	-	-	-	1/2	1/2	1/2
Israel	1/2	1	2	1/2	1/2	1/2	1/2	1/2
Japan	2	-	-	-	-	2	2	2
Korea	1	2	1	-	-	2	1/2	2
Latvia	2	2	-	-	-	-	1/2	2
Mexico	1/2	1/2	1/2	2	2	2	-	1/2
Netherlands	1/2	-	2	2	2	1/2	2	-
New Zealand[2]	1/2	1/2	2	2	1/2	1/2	1/2	1/2
Norway	2	2	-	-	1/2	1/2	-	1/2
Philippines[1]	-	-	-	-	-	2	-	-

Portugal	1/2	2	1	1/2	1/2	1/2
Romania	-	-	-	-	-	-
Russian Federation	2	2	-	2	2	-
Scotland	-	-	1/2	-	2	-
Singapore	1/2	2	-	1/2	2	1/2
Slovak Republic	2	2	2	2	2	-
Slovenia	1/2	1/2	1/2	1/2	1/2	1/2
South Africa	1/2	2	1	1/2	1/2	1/2
Spain	1/2	1	-	1	-	-
Sweden[2]	-	-	-	-	-	-
Switzerland	1/2	-	1	1/2	1/2	1/2
USA	1/2	1	1/2	1	1/2	1/2

[1] Curriculum guides submitted only for the upper grade of Population 2.

[2] See footnotes 1 and 2 in figure 4.1.

Missing Data: Argentina, Bulgaria, People's Republic of China, Cyprus, Iran (no official guides exist), Italy, Lithuania, and Thailand.

Legend:
1   Present in the curriculum guide for the upper grade of Population 1
2   Present in the curriculum guide for the upper grade of Population 2
-   Not present in either

Figure 8.7
Variation of Textbook Use of Perspectives - Population 1.
The textbook block percentages for various perspectives are shown for the upper grade of Population 1. The data for all perspectives, and for all countries, are shown. A small number of countries made considerable use of perspectives.

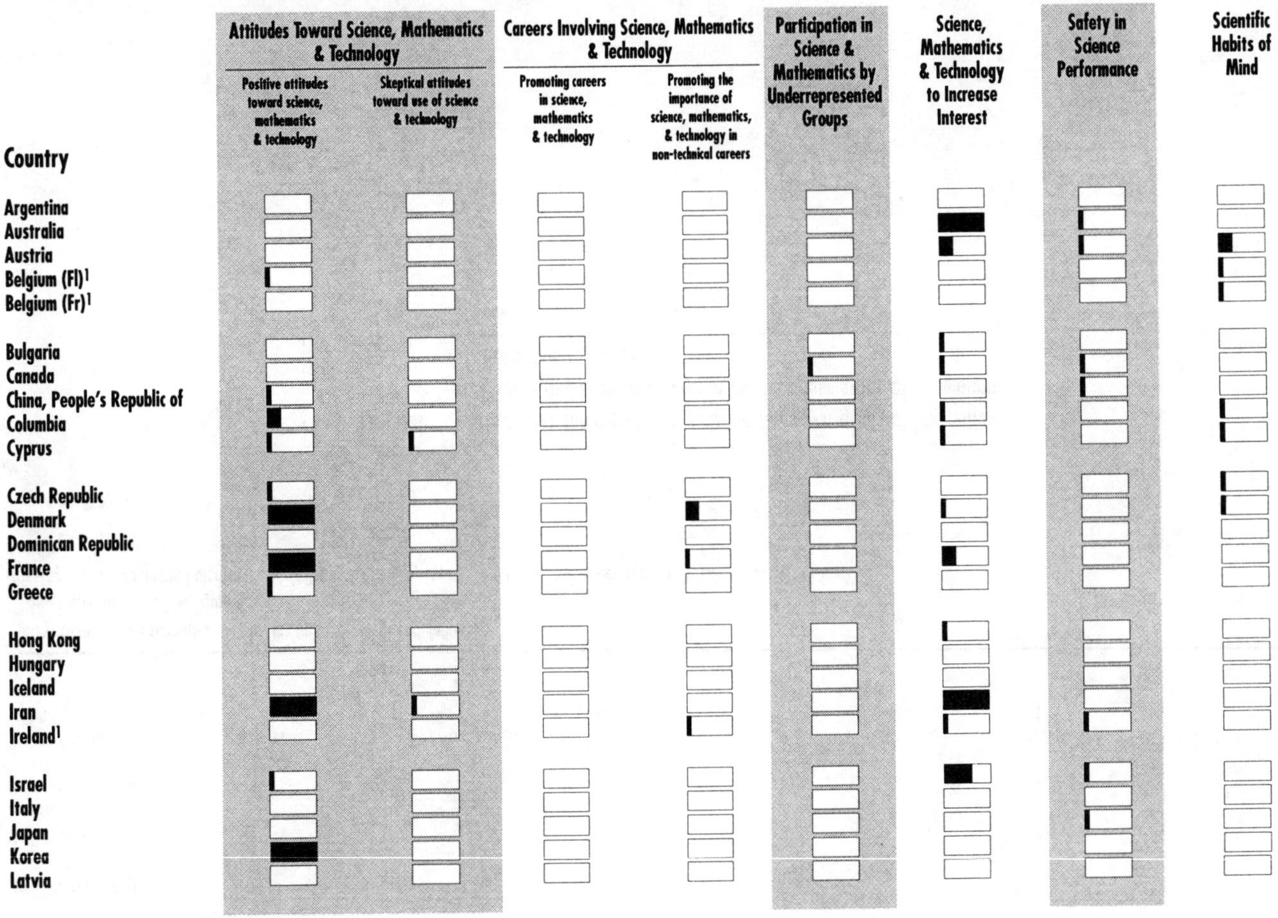

Legend:

0%	6-10%	16-20%	
1-5%	11-15%	>20%	

Mexico
Netherlands
New Zealand[1]
Norway
Philippines

Portugal
Romania
Russian Federation
Scotland[1]
Singapore

Slovak Republic
Slovenia
South Africa
Spain
Switzerland
USA

1   Refer to footnotes in figure 18.1.
Missing Data: Germany, Lithuania, Sweden, and Tunisia.

Figure 8.8
Variations in Textbook Use of Perspectives - Population 2.
Data similar to those in the previous table are shown but for the upper grade of Population 2.

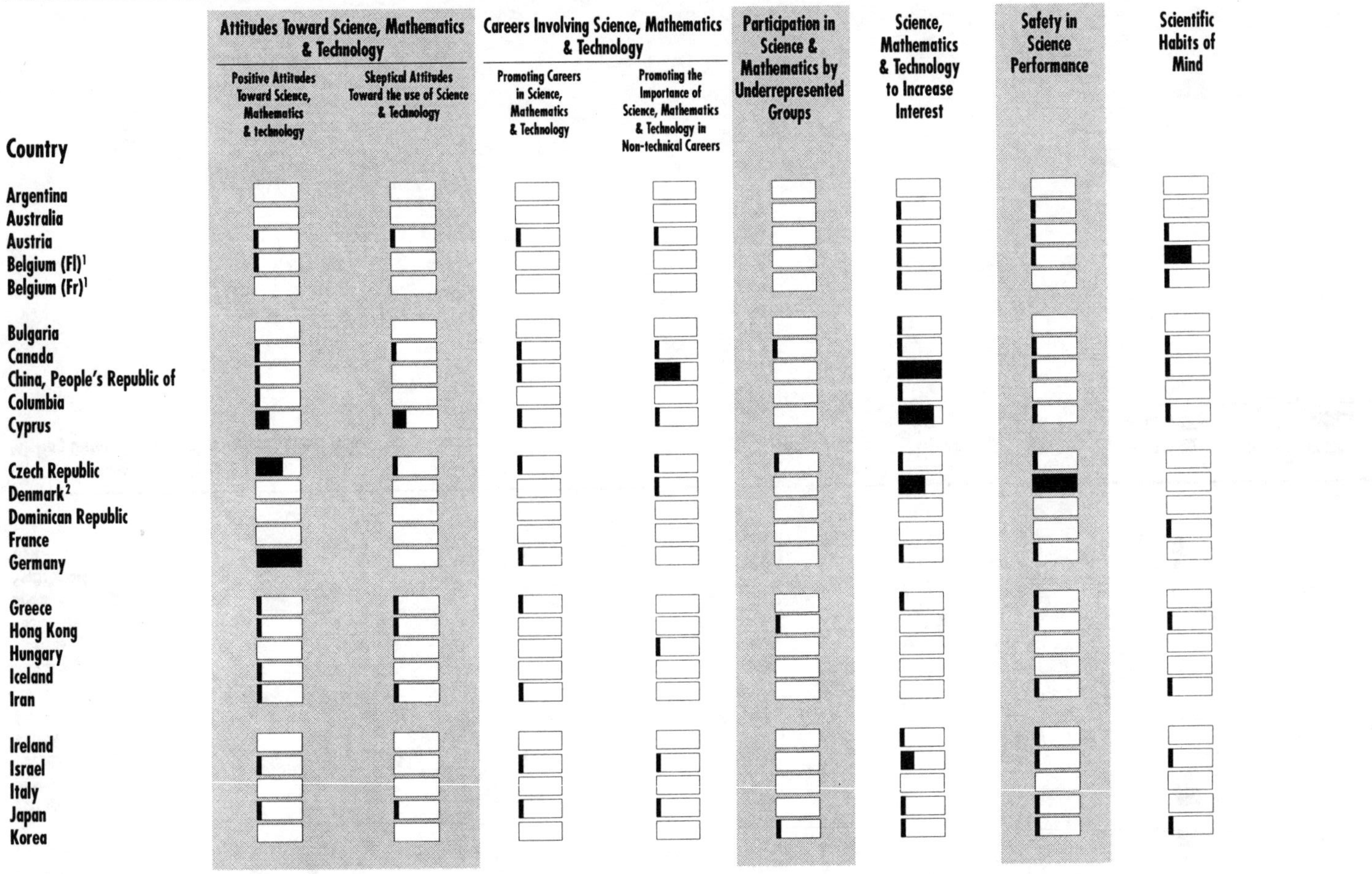

Latvia
Lithuania
Mexico
Netherlands
New Zealand

Norway
Philippines
Portugal
Romania
Russian Federation

Scotland[1]
Singapore
Slovak Republic
Slovenia
South Africa

Spain
Sweden
Switzerland
Tunisia
USA

Legend:

0%	6-10%	16-20%
1-5%	11-15%	>20%

1 Refer to footnotes in figure 18.1.
2 Denmark only provided data collected from physical science textbooks for this population.

A substantial proportion of direct statements were made in curricular documents about these matters in many countries. This finding is in marked contrast to what was found in mathematics. Some countries, such as Australia and the United States, included all of the perspective sub-categories in their curriculum guides. Other countries, for example Canada, Israel, and Tunisia used at least some of the sub-categories from all six major categories of perspectives in the science framework in both the upper grades of Population 1 and 2.

Table 8.4, figure 8.7, and figure 8.8 display the results of this analysis of perspective data. Table 8.4 covers data based on curriculum guides for the upper grades of Populations 1 and 2; the figures show textbook emphasis of perspectives. Emphasis on 'positive attitudes towards science, mathematics, and technology' was prevalent in Denmark, France, Latvia, Iran, and the Russian Federation in Population 1, although this emphasis practically disappears in Population 2 (the most notable exception is Germany). The perspectives present in textbooks intended for the upper grade of Population 2 shift mostly to 'science, mathematics, and technology to increase interest.' Latvia and, to a lesser extent, Flemish-speaking Belgium and the Philippines are notable for their inclusion of a higher proportion of blocks devoted to 'scientific and mathematical habits of mind' than was the case in other countries. In South Africa, this perspective was emphasized in the upper grade of Population 1 more than in other countries. In the case of the Philippines, higher relative emphasis on this performance expectation was a characteristic of textbooks for the upper grades of both populations.

NOTES

[1] 'Understanding simple information' is a category from the performance expectation aspect of the TIMSS science framework that refers to understanding information such as vocabulary, facts, equations, and simple concepts. Examples of performances of this type are defining, describing, naming, quoting, reciting, etc.

[2] The framework category 'interpreting data' includes extrapolating or interpolating data from a table or graph, identifying patterns or trends in data, etc.

[3] The category 'understanding complex information' refers to understanding information involving the integration of bits of simple information. Examples of performance that could indicate complex information are differentiating, comparing, contrasting, and synthesizing.

Chapter 9

RE-EXAMINING SHARED VISIONS: COUNTRY GROUPINGS

The preceding chapters have looked at curricular intentions across all participating countries. From this analysis, we see broad-based commonalities and diversities. These findings, however, mask myriad similarities and differences among countries. In an effort to delineate some of them, this chapter examines data for various groupings of countries. The patterns that emerge from these groupings yield interesting insights regarding shared curricular visions.

In undertaking this analysis, the central decision involved determining the sets of countries to investigate. Some possibilities were obvious — for example, examining the countries of various geographic regions, various economic categories, and various pre-existing market confederations such as the G-7 countries, the countries of the Organization for Economic Co-operation and Development, and European Union countries, etc. A radically different approach was to start not with groups of countries defined by pre-determined criteria, but rather to use statistical methods to cluster countries into groups and then to examine commonalities among them. This last approach could reveal interesting similarities in curricular intentions across unexpected groups of countries. All four of these grouping methods: geographic, economic, market-based, and statistical were applied in this analysis.

In the interests of space and time, this first analysis focused only on content in the narrow sense. Specifically, only textbook coverage of science framework topics was considered, as measured by the percentage of textbook blocks devoted to various content topics was considered for the upper grades of Populations 1 and 2. Other interesting approaches, including consideration of the physics specialists of Population 3, are deferred to later reports. While the data selected have certain limitations (discussed earlier), their variability make them more useful than other data (such as curriculum guide data). Obviously, what follows should be considered suggestive of, rather than conclusive for, inferences about curricular visions and intentions.

Countries Grouped by Geographic Region

The most obvious criterion for grouping countries is geographic region. Neighboring or nearby countries often share traditions, cultural similarities, and historical interactions. They may have a history of — or an interest in — mutual comparisons, or may have a more direct economic competitiveness. They also might share common curricular traditions.

There are many ways of assigning regional categories to the various countries included in this study. The grouping used in this analysis is based on a recognition that participation in the Third International Mathematics and Science Study (TIMSS) was not uniform across the world. This grouping[1] comprises eight regions as generally determined by geographic, historic, and cultural proximity:

- Latin America (Spanish-speaking);

- Australia and New Zealand;

- East Asia;

- Western Europe;

- East and Central Europe;

- United States and Canada;

- North Africa and the Middle East; and

- South Africa.[2]

Table 9.1 shows the average percentage of textbook blocks devoted to each of the intermediate categories of the science framework for the upper grade of Population 1 by the geographical regions outlined above. Table 9.2 does the same for the upper grade of Population 2.

One area of difference was in the attention devoted to 'earth features' topics — a central group of topics in Population 1 — and to sub-topics in the life sciences. The textbooks of countries grouped as East and Central Europe indicated almost twice the textbook coverage of 'earth features' at this level than did those of the United States and Canada and Western Europe, and both of these regions devoted considerably more attention to this topic than the others.

The group of Middle East/North African countries were remarkable in the extent to which the physical sciences were emphasized, as were Australia and New Zealand. Topics such as those included in 'energy and physical processes' were not emphasized as much in the other regions.

There were other variations for the upper grade of Population 2 (see table 9.2). Canada and the United States and South Africa emphasized 'earth features' in the upper grade of Population 2 more than other regions. Life sciences were emphasized by the Latin American and Middle East/North African regions, with Latin America devoting more attention to 'diversity, organization, structure of living things' than other regions.

There are interesting regional variations in the emphasis on physical science topics. In Latin America, much more emphasis is given to topics in 'forces and motion' than in any of the other regions. 'Energy and physical processes' is emphasized more in the textbooks of East and Central Europe.

Table 9.1
Topic Coverage in Eight Regions — Population 1.

These data are the average percentage of textbook blocks for each science topic averaged over the countries in one of eight regions. They are only for the upper grade of Population 1. Notable variations among regions occur in the areas of 'energy and physical processes,' 'diversity, organization, structure of living things,' and 'earth features.'

Science Topics	Latin America	Australia and New Zealand	East Asia	Western Europe	East and Central Europe	U.S. and Canada	Middle East/ N. Africa	South Africa
Earth Sciences								
Earth Features	13	2	6	15	28	16	4	4
Life Sciences								
Diversity, Organization, Structure of Living Things	39	29	41	36	45	42	1	42
Interactions of Living Things	17	10	7	10	12	9	0	17
Human Biology and Health	5	0	13	8	5	7	7	2
Physical Sciences								
Energy and Physical Processes	11	43	17	14	8	20	53	0
Forces and Motion	3	17	0	4	6	5	0	0
Environmental/Resource Issues								
Food Production, Storage	7	0	1	3	2	1	1	12

Table 9.2
Topic Coverage in Eight Regions — Population 2.

These data are similar to those in the previous table but are only for the upper grade of Population 2. In addition to regional variations on some of the same topic areas as noted for Population 1 in table 9.1, a variety of earth, life, and physical science topics present noteworthy regional variation.

Science Topics	Latin America	Australia and New Zealand	East Asia	Western Europe	East and Central Europe	U.S. and Canada	Middle East/ N. Africa	South Africa
Earth Sciences								
Earth Features	6	9	12	5	7	21	3	17
Life Sciences								
Diversity, Organization, Structure of Living Things	50	23	13	17	13	18	18	6
Life Spirals, Genetic Continuity, Diversity	14	3	3	6	6	9	37	21
Interactions of Living Things	13	9	12	4	2	10	9	6
Human Biology and Health	4	12	4	12	8	6	4	0
Physical Sciences								
Matter	15	13	11	10	18	18	9	13
Structure of Matter	5	5	5	2	9	4	4	8
Energy and Physical Processes	19	25	21	27	34	21	18	9
Physical Transformations	5	7	3	4	4	8	2	3
Chemical Transformations	4	9	8	7	13	5	7	1
Forces and Motion	20	9	3	7	4	5	4	1
Science, Technology, and Mathematics								
Interactions of Science, Mathematics, and Technology	1	2	4	4	3	12	1	0

Countries Grouped by Economic Category

Economic considerations might have some impact on science curricula. To discern this impact, countries were grouped by average income. Using the World Bank's *World Tables 1995*, country incomes were categorized in four ways:

- *low* — having a per capita gross national product (GNP) of US$695 or less in 1993;

- *lower middle* — a per capita GNP between US$695 and US$2,785 in 1993;

- *upper middle* — a per capita GNP between US$2,785 and US$8,626; or

- *high* — a per capita GNP greater than $US8,626[3].

The low-income group was represented by only one country — the People's Republic of China. Thirteen countries were represented in the lower middle category, 8 in the upper middle, and 23 in the high-income category. Except for the low-income group, this representation seems sufficient for substantive contrasts. Notice that per capita GNP is used here only as an indication of comparative wealth and not as a direct indicator of educational opportunity.

Table 9.3 presents the average percentage of textbook blocks devoted to each of the science framework topics for the upper grade of Population 1 for the four income categories outlined above. The most striking observation here is how few distinctions exist at the Population 1 level. Among these countries, the high-income group did less with 'earth features' than the middle-income groups. The upper middle-income group did less with 'energy and physical processes' than was the case in the other middle-income group and the high-income group.

There were more variations in topic coverage by income group for the upper grade of Population 2 (see table 9.4). The high-income group paid somewhat less attention to 'structure of matter' than the lower middle-income group. 'Energy and physical processes' received more attention from the lower middle-income group.

In some ways, the results of these analyses are surprising. It would appear that grouping countries on geographic criteria uncovers more striking contrasts than grouping them by income. Although it may be our hope that income not affect the quality and variety of educational opportunities that countries can make available to students, other evidence suggests that this is not true. It is a simple fact in science education that some pedagogical strategies are more expensive than others, a fact that has been recognized for some time by organizations such as the World Bank.

By the results reported here, however, this hard truth is not evident in topics covered. However, future analyses will examine this issue from the perspective of performance expectations and block types. Practical activities in the sciences are particularly expensive, for example, which implies they might be used with more or less frequency in different countries according to their economic circumstances. Future work will attempt to discover whether the intended curriculum of TIMSS countries vary in performance expectations for science according to

Table 9.3
Topic Coverage for Four National Income Groups — Population 1.

These data are the average percentages of textbook blocks for each science topic averaged over the countries in four national income groups (determined by World Bank criteria and data). They are for the upper grade of Population 1 only.

Science Topics	Low	Lower Middle	Upper Middle	High
Earth Sciences				
Earth Features	3	23	22	9
Life Sciences				
Diversity, Organization, Structure of Living Things	18	38	39	38
Interactions of Living Things	5	12	12	9
Human Biology and Health	0	7	5	9
Physical Sciences				
Energy and Physical Processes	2	16	6	20
Forces and Motion	0	3	2	5
Environmental/Resource Issues				
Food Production, Storage	0	4	3	2

Table 9.4
Topic Coverage for Four National Income Groups — Population 2.

These data are similar to those for the previous table, but are only for the upper grade of Population 2.

Science Topics	Low	Lower Middle	Upper Middle	High
Earth Sciences				
Earth Features	31	7	8	7
Life Sciences				
Diversity, Organization, Structure of Living Things	25	25	13	17
Life Spirals, Genetic Continuity, Diversity	9	10	10	8
Interactions of Living Things	6	6	4	7
Human Biology and Health	1	5	12	9
Physical Sciences				
Matter	2	15	12	12
Structure of Matter	0	7	6	3
Energy and Physical Processes	2	31	16	27
Physical Transformations	0	5	3	4
Chemical Transformations	0	10	9	7
Forces and Motion	11	5	12	6
Science, Technology, and Mathematics				
Interactions of Mathematics, Science, and Technology	5	3	4	4

their economic circumstances — reflecting the need to limit particularly costly pedagogic strategies in countries with less resources.

Countries Grouped by Market Group Category

A third approach to grouping countries was by "market groups" of close economic blocks. These groups may not have always shared common traditions or regional similarities, but they do have a special interest in cross-national comparisons of educational productivity and other related educational system factors of the countries of their market group. One such grouping is considered here — the European Union (EU) countries versus other countries. Of the 15 current EU members, 12 are represented in these data; only the United Kingdom, Finland, and Luxembourg are not.

Differences in textbook coverage of science framework topics for the upper grade of Population 1 in EU countries versus non-EU countries were few. On average, EU countries had less textbook content in 'energy and physical processes' topics. They did more with 'diversity, organization, structure of living things.'

The differences for the upper grade of Population 2 were neither particularly numerous or strong. The other countries did somewhat more with 'life spirals, genetic continuity, diversity' than did the EU countries.

Countries Grouped Statistically

A quite different approach to investigating similarities in limited country groups was to cluster countries by their data. Here the country clusters were likely to be more diverse than the other groupings; in fact, they might only have some content emphases in common. For this reason, the investigation focused primarily on which topics formed the basis for similarities. This focus differs somewhat from that of the preceding analyses; there, we looked at both the degree and type (content) of differences; here, attention is focused solely on the content sources of differences.

Based on separate statistical clustering of Population 1 and Population 2 (upper grade) textbook data, two sets of clusters (one set for each population) emerged that shared major similarities in content coverage. In addition, certain countries among each population — Cyprus, Iceland, Iran, Israel, and the Philippines in Population 1; Denmark, the Dominican Republic, and Israel in Population 2 — revealed such distinctive content patterns that it seemed better not to group them in any of the clusters. (It is worth noting that mathematics curricula in the Dominican Republic and Israel were also so distinctive that they were excluded from the clusters made for the analyses of mathematics curricula.)

For the upper grade of Population 1, strong differences were seen only for a few topics (see table 9.5). 'Land forms' and 'bodies of water' separated the clusters into groups, as did relative

Table 9.5

Topic Coverage for Eight Statistically Determined Clusters — Population 1.

These data are the average percentages of textbook blocks for each science topic area averaged for each of eight statistically determined clusters of countries. Several strong differences were seen.

Science Topics	Cluster Group[1]							
	Austria, Belgium (Fr), Mexico, Portugal	Colombia, New Zealand, Scotland	Argentina, Australia, People's Republic of China, France	Czech Republic, Greece, Japan, Korea, Romania	Canada, Slovak Republic, Hungary, Netherlands, Norway, Russian Federation, Spain	Dominican Republic, Hong Kong, Italy, Singapore, Switzerland, USA	Belgium (Fl), Bulgaria, Latvia, South Africa	Denmark, Ireland, Slovenia
Earth Sciences								
Earth Features								
Landforms	12	1	1	11	4	0	6	0
Bodies of water	8	0	1	11	5	1	5	1
Life Sciences								
Diversity, Organization, Structure of Living Things								
Plants, fungi	4	8	3	15	17	16	20	27
Animals,	3	4	4	11	17	15	19	47
Organs, tissues	4	2	13	3	10	26	3	7
Interactions of Living Things								
Biomes and ecosystems	2	5	4	1	5	1	3	3
Habitats and niches	2	8	0	1	3	1	2	4
Human Biology and Health								
Nutrition	0	0	5	3	5	9	0	1
Disease	0	1	2	0	3	3	1	0
Physical Sciences								
Matter								
Physical properties	2	14	6	6	3	5	2	5
Energy and Physical Processes								
Electricity	2	13	3	5	2	3	0	0
Forces and Motion								
Types of forces	1	7	3	1	3	1	0	1

[1] Countries not included in this table are Cyprus, Germany, Iceland, Iran, Israel, and Philippines.

Table 9.6

Topic Coverage for Eight Statistically Determined Clusters — Population 2.

These data are the average percentages of textbook blocks for each science topic averaged for eight statistically determined clusters of countries. The data are only for the upper grade of Population 2.

Science Topics	Cluster Group[1]							
	Argentina Cyprus Latvia Philippines	Canada Mexico Netherlands Norway Spain	Colombia Tunisia	Bulgaria People's Republic of China Greece Iceland New Zealand USA	Austria Belgium (Fr) Hong Kong Hungary Italy Portugal Singapore Sweden	Iran South Africa	Australia Belgium (Fl) Ireland Lithuania Russian Federation Scotland Switzerland	Czech Republic France Japan Korea Romania Slovak Republic Slovenia
Earth Sciences								
Earth Processes								
Weather & Climate	1	6	2	2	2	12	1	3
Earth in the Universe								
Earth in the Solar System	2	2	1	2	2	7	1	0
Life Sciences								
Diversity, Organization, Structure of Living Things								
Animals	0	2	4	7	4	0	12	2
Organs, Tissues	3	5	5	11	6	2	9	3
Life Processes & Systems Enabling Life Functions								
Energy Handling	3	3	11	4	4	0	4	3
Life Spirals, Genetic Continuity, Diversity								
Life Cycles	1	1	13	3	0	9	3	0
Reproduction	3	2	16	2	2	10	2	2
Interactions of Living Things								
Animal Behavior	0	1	8	2	1	0	1	1
Human Biology & Health	1	3	0	11	4	0	9	4
Physical Sciences								
Matter								
Classification of Matter	2	5	8	2	4	3	6	4

Energy and Physical Processes								
Energy Types, Sources, Conversions	14	4	2	6	3	8	5	3
Heat & Temperature	6	3	1	2	4	0	6	1
Light	7	1	0	3	6	5	4	2
Electricity	0	1	0	1	8	9	10	23
Physical Transformations								
Describe Physical Changes of Matter	3	3	6	4	1	4	2	1
Chemical Transformations								
Describe Chemical Changes of Matter	2	3	1	3	4	3	3	8
Forces & Motion								
Time, Space, & Motion	4	3	2	3	2	0	1	1
Dynamics of Motion	3	1	2	1	1	4	1	0
Science, Technology, & Mathematics								
Interactions of Science, Mathematics, & Technology								
Applications of Science in Mathematics, Technology	6	2	0	5	4	0	3	3
Environmental/Resource Issues								
Food Production, Storage	0	2	8	2	1	0	2	1

[1] Countries not included in table: Denmark, Dominican Republic, Germany (did not code Biology text), and Israel.

emphasis on topics related to 'diversity, organization, structure of living things,' 'interactions of living things,' and some topics in the physical sciences. The cluster of Colombia, New Zealand, and Scotland, for example, was distinguished by much attention to 'electricity' and 'physical properties of matter,' and less attention to 'bodies of water,' 'human biology and health,' and 'organs, tissues' at this level.

A more diverse set of topics formed the basis for clusters in Population 2, as seen in table 9.6. 'Energy types, sources, conversions' helped separate some clusters. 'Light' and 'electricity' also made contributions to differences. 'Classification of matter' was less important in this regard. Emphasis on 'weather and climate' and on 'earth in the solar system' helped to distinguish the cluster formed by Iran and South Africa. Degree of emphasis on 'electricity' was the single most distinguishing feature of the cluster of countries formed by the Czech and Slovak Republics, France, Japan, Korea, Romania, and Slovenia.

The statistical clustering of countries created interesting groups for further exploration of similarities. At first glance, these groups appeared quite disparate, and their clustering was not easy to interpret. However, what clearly emerges from this approach is that there are large differences among clusters on key topics. The magnitude of these differences prove that there were strong intercountry differences on these topics — the pooled intercountry variance on a few topics was sufficient for groups of countries to be easily distinguished. In this context, it is noteworthy that the traditional basis for grouping — geographic region, economic factors, market and political affiliations — revealed cross-national differences only by an accumulation of many diverse, yet small, topic differences. The traditional a priori groupings tried here did not uncover key topic variances of magnitude in distinguishing groups. Statistical approaches did *not* reify a priori groupings — traditional factors simply were not as effective in revealing content distinctions in this case.

NOTES

[1] Countries were assigned to the following regions: Latin America (ARG, COL, DRP, MEX); Australia and New Zealand; East Asia (CHN, HKG, JPN, KOR, PHL, SGP, THA); Western Europe (OST, BFL, BFR, CYP, DNK, FRA, DEU, GRC, IRL, ISL, ITA, NLD, NOR, PRT, SCO, ESP, SWE, SUI); East and Central Europe (BGR, CSK, HNG, LVA, LTU, ROM, RUS, SVK, SVN); the United States and Canada; South Africa; and North Africa and Middle East (IRN, ISR, TUN). See appendix F for a list of the country abbreviations.

[2] The composition of these latter two regions was dictated primarily by the low representation of African and Middle Eastern countries in TIMSS.

[3] Countries were assigned to the following groups in accordance with the "classification of economics by income" in the 1995 issue of the World Bank's *World Tables:* Low Income (CHN); Lower Middle Income (PHL, THA, BGR, CSK, LVA, LTU, ROM, RUS, SLV, IRN, TUN, COL, DRP); Upper Middle Income (SAF, KOR, HNG, SVN, GRC, PRT, ARG, MEX); and High Income (AUS, JPN, NZL, HKG, SGP, OST, BFL, BFR, DNK, FRA, DEU, ISL, IRL, ITA, NLD, NOR, ESP, SWE, SUI, CYP, ISL, CAN, USA). See appendix F for a list of the country abbreviations.

IV. CONSEQUENCES OF CURRICULUM: POLICY IMPLICATIONS

Chapter 10

CONCLUDING REMARKS

The analysis presented in this monograph represents a first attempt to examine and compare the curricular aims of the countries participating in the Third International Mathematics and Science Study (TIMSS). Several pervasive conclusions were identified during this effort and they are summarized in this chapter. Also, since the curriculum data gathered are complex and the results presented here are only from a first set of investigations, we outline some directions for future curriculum analysis work.

A Few Broad Conclusions

Pervasive Variation

The preceding chapters repeatedly presented examples of pervasive and detailed cross-national variation. Variation was pervasive, occurring among countries and topics in myriad ways both small and large. Commonalities always occurred against this background of variation among topics and countries. Variation was also detailed, occurring at virtually every level of detail. Important commonalities — both broad and, at times, specific — were found by examining topics across countries, countries across topics, or both. However, at whatever level commonalities were found, the details at that level always varied both among countries and among topics.

While it is perhaps premature to generalize from the TIMSS data, it seems unlikely that data gathered from detailed, empirical analyses of curricular materials, educational materials, or cross-national testing will ever be found free of similar detailed, pervasive variation. This climate of variation is the climate of large-scale empirical investigations into aspects of national education systems and their activities. A study that does not find or does not take into account such variation seems suspect and would seem to merit careful discussion of why such variation did not occur or was not considered.

The Importance of Context in Interpreting Findings

Cross-national educational comparisons are based on finding significant similarities and significant differences. While it is tempting to see detailed variation as "noise" and only broad similarities and major differences as "signal" or "message," such distinctions may be difficult — if not impossible — to make when data have been oversimplified or their context has been ignored.

All comparative data must, therefore, be presented within a context of variation. This context must be carried throughout the analysis and summarized appropriately when stating findings. Cross-national comparisons of educational systems and their science education activities are intended to provide information on curricula, educational systems, national goals, and how well national goals are being met. Analytic comparisons are only meaningful if they are based on factors sensitive to the effects of curriculum and educational systems, and relevant to national goals in science education. Sensitive comparisons considered apart from this context are unnecessarily simplified and the validity of these results is necessarily suspect.

Weaknesses of A Priori Comparisons

Comparative educational research has often uncovered important educational differences between countries grouped according to geographic, economic, or other criteria. However, such a priori groups only served to uncover *some* differences in the curricular factors investigated here.

The groups formed by rational, a priori criteria (which were discussed in chapter 9) differed primarily by the cumulative effect of many small differences among content topics and curricular emphases, rather than marked differences among a few topics. It is here that the statistical clustering reported has a special significance. Such statistical clustering, by the very nature of the analysis employed, maximizes the distinctions among groups of countries. The topics identified by such analyses and the size of the differences among the clusters are essentially as large as can be obtained from the variation that exists in the data.

In this sense, the statistical clustering suggests two conclusions:

- There are some topics where marked variation can be found when the basis for forming groups of nations is allowed to range freely to take into account as much as possible the cross-national variation that occurs. However, the national groups identified in this way bear little resemblance to the a priori groups.

- Grouping nations by those a priori criteria considered was less effective in revealing the meaningful structure of similarities and differences inherent in the variation of the original data. This may be true because the most appropriate a priori criteria have yet to be identified and investigated. Nonetheless, it may well be that the groups formed by a priori logical or rational criteria for cross-national comparisons of science curricula may provide only for the simple accumulation of small differences. In this case, further study — rather than further attempts at alternative grouping criteria — is needed to explain more meaningful and substantive similarities and differences.

Possibilities and Limitations of Large-Scale Document Analysis

The TIMSS curriculum analysis effort and, in particular, its document-based data collection and analyses, was not merely a feasibility study for such cross-national analyses on a large scale. These activities were a serious first attempt at large-scale document analysis and were intended to yield useful results. However, they were also a feasibility study of the possibilities and limits of large-scale curriculum document collection and analysis.

Large-scale cross-national curriculum comparisons previously have relied mostly on surveys and expert data. Analysis of curricular documents has typically been reserved for smaller scale studies with limited comparisons, often those with a case study approach, as much qualitative as quantitative. The present investigations suggest that document-based data gathering and analyses are feasible on a large scale and can yield useful data. Revealing data can be collected without translation of entire documents. The data collection can be sufficiently detailed for quantitative analyses. While case studies are useful sub-analyses even in the context of this larger scale approach, analysis of curricular documents is not inherently limited to case studies.

On the other hand, these investigations have shown how demanding such research is. The cost in time, personnel, and other resources certainly increases more than linearly with increases in the scope and scale of curriculum document collections and analyses. The sort of statistical abstracts of documents seemingly necessary for document analysis on such a scale requires careful prior design for (1) data collection constructs and methods, since the abstracting process will limit analytic possibilities, and (2) procedures to ensure and assess data reliability and validity. The decisions made in planning the TIMSS collection permitted certain kinds of analyses and eliminated others. TIMSS also showed some of the magnitude involved in data quality assurance and in assessing reliability and validity in the analysis of documents coded by national teams without full translation into a common language.

Future Directions

Future Planned Investigations

The TIMSS curriculum data are tremendously detailed and extensive; this monograph reports the very first steps of the analyses that can and will be performed using these data. For example, planned investigations will cover the structure of textbooks and the relation of textbook form and function. Also, future monographs will report more fully on the science curriculum and on the relation of the science and mathematics curricula.

As the full sets of TIMSS curriculum data, survey-based data (for schools, teachers, and students), and achievement test results become available, analyses will be made to relate intention, implementation, and educational attainments. The initial focus will be on cross-national comparisons. However, the TIMSS curriculum data have shown how extensive intracountry variation is among participating countries. Further studies are needed to investigate intracountry variations in nations with regional and other sub-national educational systems and varying science curricula. Also, studies of nation group comparisons — as opposed to comparisons across the whole set of countries — have barely begun. Further investigation is needed of limited country groups to find substantive similarities related to national and educational system factors.

Additional Data Needs

Although ambitious, the TIMSS data collection effort revealed important omissions in its design resulting from necessary tradeoffs and resource limitations in planning the collection. To address these omissions and enhance the value of the existing TIMSS data, at least small-scale supplemental data collections are needed. Examples of information that would fill in gaps and be highly useful in future TIMSS analyses include the following:

- Direct information on sequence and topic emphasis;

- Documents other than curriculum guides and textbooks — for example, annotated teacher textbooks and national examinations;

- Determination of the differing role and function of textbooks among nations and cross-national differences in that role; and

- More detailed information about curriculum guides.

Secondary Analyses of Existing Data

Mechanisms are in place for more extensive analyses of the existing TIMSS data, especially the curriculum data. These primary analyses will certainly not exhaust the possibilities of the data and their interrelationships. As time and resources permit, the data will — upon completion of the primary analyses — be made available for secondary analyses. Documentation must be provided for the extensive database, as well as provisions made for sharing a data collection of this massive size. The corresponding original curriculum documents are also archived and will be made available for future analyses.

New Methodologies

The volume and complexity of the TIMSS data challenge the existing state of the art in educational research methodologies. They point up the need for newer analytical methods. We have here applied one set of curriculum analysis methods to this large data collection. Other newer, and more penetrating, analytic methods should provide additional insights.

A Final Remark

The findings and conclusions presented in this monograph are only a beginning. More is planned for primary analyses of the TIMSS data; more still should result both from eventual secondary analyses and further data collections suggested by these analyses. The purpose of this report is both to answer some persistent questions and to raise some new questions.

V. Supporting Data

Appendix A

CURRICULUM, EXPERIENCE, AND OPPORTUNITY

This appendix details the theory — presented briefly in chapter 1 — of curriculum and its relation to actual student experiences. This theory informs the view of curriculum used throughout most of the analyses presented in this monograph.

Experience and Intentionality

Directly shaping students' educational experience is to some degree problematic. Given this reality, does it even make sense to speak of curricular intentions — of guiding visions and aims shaping curricular activities? Common sense and educational practice certainly say "yes." Even with students' experiences as curriculum building blocks, there is room for "builders" — those who act deliberately and with intention — to affect those experiences. There must always remain a fundamental sense in which individual experience is not malleable. However, if that were the final story, there would be no room for the study of curriculum or the practice of schooling.

The history of goal-driven group educational attainments makes it clear that much of what guides educational experience is alterable. For a category of students (participants in a specific educational system, of a similar age, and so on), similar opportunities (backgrounds, educational settings, tasks, activities, materials, and involvement) typically lead to similar educational attainments. In specific situations, educational attainments are comparatively predictable, and the distribution of such attainments is often relatively narrow for similar students. If this is so, a reasonable inference is that the experiences of individual students lying behind specific attainments are shared, common experiences. Thus, given that it is sensible to speak of approximately common experiences, it seems reasonable also to examine those manipulable elements that help to produce them. These elements (activities, materials, etc.) are the creations that reflect the intentions of those who made them.

The present volume considers curriculum only in its more narrow, policy-malleable aspects. Curriculum is intentional and related to students' experiences in schooling. The aims and intentions of policy- and curriculum-makers articulate visions and seek to guide experiences. They impose authority and "[bring] order to the conduct of schooling."[1] This volume is, then, in part a study of how curriculum orders students' experiences in schooling and of the structures imposed by formal authorities in a variety of national educational systems. It does not examine the direct mechanisms for imposing such order — a task left for later volumes reporting this study. Rather, it examines some sets of curricular data and materials (curriculum guides and textbooks) for patterns of reflected intentions and for commonalties and diversities of order reflected in materials from various educational systems. It also investigates what may be inferred about common and differing intentions.

Opportunity and Potential Experience

Further clarification of the relationship between curricular experiences and educational opportunity may be needed. Students' experiences are here considered as a matter of individuals' interior lives. Teachers may help create and shape experiences, but students' actual experiences are combinations of activity and setting — of opportunity or "potential experiences" — and of children's participation and engagement in those activities. This combination of opportunity, engagement, and those engaged determines the actual experiences flowing from encounters with potential experiences. Those guiding educational systems plan the creation and distribution of educational opportunities. Teachers shape specific opportunities — settings, activities, and their short-term sequences. Students reshape opportunities by their perceptions and reactions — acting as the final arbiters in creating their own experiences. It is out of those actual experiences that learning occurs, and it is this learning that achievement tests measure.

Remembering this, curriculum should not be seen as directly managing and ordering students' experiences, especially when educators are considered to express intentions through curriculum guides and textbooks (and other curricular materials). Articulating visions and aims and planning the creation and distribution of desired common potential experiences are only the first stages of the varied tasks of devising curriculum. Later stages must follow — steps taken by teachers and students. One may judge the outcome of the first phases only by considering the entire result — both the vision and final outcomes. Even as the curriculum policymakers map the domains, topics, performances, etc., of intended experiences, providing intended sequences of those experiences, students must interact with those intentions as teachers implement and shape them. Many factors help determine how similar the curriculum as experienced by students is to the intentions of educational policymakers. It depends at least on the detail and clarity of the goals and plans, on the difficulties that the system inadvertently creates along the way, and on the resources available to teachers and students. While this study is about how national intentions are communicated and used to help guide students through the course of their studies, acts of curricular intentionality are worth investigation in themselves. As they help to determine the futures of students, these first stages — these goals and sequences of intended student experiences — *must* be investigated.

One must remember the indirect nature by which "desired" common experiences are sought in the pursuit of educational aims and curricular intentions. This seems best done by use of a more indirect term — such as *opportunity* — for that which is directly manipulated by educational policies and by curricular materials (especially, curriculum guides and textbooks). *Opportunity* is used only to maintain in the foreground this essential indirectness in shaping students' experiences. It is an expression of "desired" and "potential" experiences that, in the end, can be shaped only partly but not determined completely.

Opportunity for some is a highly polemical word that, when coupled with the word *education*, often has political connotations associated with a complex set of values regarding equality and other social norms. Some consider any statement that implies a failure to provide an opportunity as an indictment of educational practice. However, since students' time for schooling is limited, which, in turn, limits the numbers of activities and settings, educational opportunity also is limited — in economic terms, it is a "scarce" resource. Every choice shaping schooling provides some opportunities at the expense of others. The opening tasks of curriculum-making involve using vision and insight to define, choose, and sequence opportunities. National choices are clearly the concerns of individual nations (and systems) — matters of their particular visions — considered as made with the intention of serving well both the nation and its children.

Common opportunities — potential experiences — rather than common experiences, are a much more conservative basis for discussing curricular aspects that can be "charted" — intended, guided, and ordered. This volume considers curriculum only in its role of distributing potential experiences and intending to aid the management of sequences, patterns, and arrays of opportunities. It uses only language of description, not of judgment. Those wishing a more detailed look at this volume's underlying model of educational opportunities as potential experiences should look at appendix B.

The visions behind science curricula have many aspects: conceptions of science, perceptions of students, perceived national needs for educational outcomes, and so on. One central aspect of science curricular visions is determining which opportunities for science experiences should be provided to students at particular points in schooling — opportunities intended to provide a core of common science experiences and outcomes. The focus of investigation is on aims; the data sources are mainly selected documents and the main aspect of national aims to be illuminated are the intended opportunities for common science experiences.

Curriculum and the International Association for the Evaluation of Educational Achievement (IEA)

TIMSS is an IEA study, and these studies — since the late 1970s — have considered curriculum as a complex construct with several facets, each linked to a context or level of educational activity. Curriculum is considered as *intention* — seen in national policies and official documents reflecting societal visions, educational planning, and official or political sanctioning for educational objectives. At the level of teacher and classroom activity, curriculum is considered as *creative implementation* of those intentions and objectives. At the level of student outcomes, the curriculum is considered as *attainment* — the result of what takes place in classrooms. Academic achievement and student belief measures document part of these student attainments.

A commonplace assumption of educational researchers has long been that how students perform on tests is closely related to what they experience in classrooms. These opportunities for experience — the instructional events and activities in which students have participated and that help prepare them to solve test items — have been conceptualized as "opportunity to learn" (OTL) in past IEA and other studies. The OTL concept has been measured typically by having teachers examine tests to be administered to their students. Teachers then report whether the students have been taught the skills necessary to complete the tested tasks satisfactorily. The index for OTL often has been simply the percentage of teachers in a country reporting that students have been given adequate opportunity to answer specific test items. Several variations of the questions prompting such teacher ratings have been used in past studies.

Previous IEA studies have suggested that this conception of educational opportunity as OTL — a measure based on teacher ratings of individual achievement test items — has serious faults. While a variant of traditional OTL is used in TIMSS, from the first, the intention has been to use a more sophisticated conception of educational opportunity (see appendix B) when investigating science curricula — both as intention and as implementation. The curriculum analysis — for which this volume is an initial report — is part of that attempt to use a more informative concept of educational opportunity to help better characterize curricular visions and intentions. In this, it goes beyond traditional IEA OTL approaches.

Defining Curricular Intentions: Curriculum Guides and Textbooks

Curriculum guides (or whatever they are called in participating countries) articulate official policies as they apply to large groups of students — to all students in a certain grade, in a specific type of school, etc. In short, curriculum guides set out curricular intentions. Each is an official, comparatively global guiding statement for what a science curriculum is intended to be in a specific context and, perhaps, how related instruction is to be conducted. Textbook authors write at least in part to support implementation of national intentions. They reflect — with differing clarity — science curricula as set out in curriculum guides (present and past) but as interpreted by one set of individuals, the textbook authors.

Textbooks have official status in some countries, and the clarity with which they reflect official curriculum is carefully controlled. In other countries, textbooks are developed more independently — and may reflect more independent interpretations of curricula as set out in (one or more, current or past) curriculum guides. Their status as "official" documents correspondingly varies, from official curriculum documents supplementing curriculum guides to completely unofficial materials for implementing instruction that reflect independent, and even private, curricular interpretations.

Textbooks are not widely available in some countries or are available mainly for teachers to help carry out classroom activities. In many other countries, textbooks are the single, standard instructional resource most teachers and most students share. Curriculum guides almost always officially define curriculum for large groups of students. In many countries, however, students are not even aware of these documents. Often curriculum guides are not consulted daily even by teachers. By contrast, evidence suggested that textbooks were present in almost every classroom in the TIMSS countries and were used regularly in instruction. Thus, while textbooks were often not *officially* national in character, they were a common element in most classrooms in most TIMSS countries.

The official *status* of curriculum guides was comparatively clear, yet curriculum guides varied in design, function, and detail as much as the government policies they articulated. In fact, guides varied so greatly both between and within countries that what was considered a curriculum guide in one country might be unrecognizable as such in another. See chapter 4 for a further discussion of curriculum guides.

In contrast, the official status of textbooks varied greatly among the TIMSS countries, yet textbooks were instantly recognized as commonplace in classrooms in every country despite variation in design and use. Textbooks provided a more detailed map of scientific disciplines, topics, and performances to be mastered in pursuing national goals: they were the "potentially implementable" curriculum. Each was a detailed chart proposing a particular path among the aspects of a science curriculum. Certainly textbooks did not force teachers and students to follow the same path in every classroom in which they were used. They did attempt, however, to provide one strategy for navigating among science topics and skills and to support a range of varied implementations within a broad reflection of curricular intention. Figure A.1 suggests that textbooks serve as intermediaries in implementing science curricular intentions.

The ways in which curriculum guides and textbooks were used officially to state curriculum — and textbooks less officially to guide implementation — are varied, and thus using these documents to investigate curricular intentions was difficult. Methodologically, curriculum guides were considered as providing the primary — and purest — official, public statement of curriculum and curricular intentions. Textbooks were considered as providing — for most countries — a supplemental, supporting reflection of those intentions. For most countries, textbooks were viewed as a primary bridge between intention and implementation — between the ministry and the classroom. While care must be taken in interpreting textbook data — given different national status for and uses of textbooks — research conducted in many countries indicates the enormous influence of textbooks on student achievement.[2] Neither curriculum guides or textbooks could stand alone as the sole source for the TIMSS curriculum analyses.

Figure A.1
Textbooks — The Potentially Implementable Curriculum.
Textbooks served as intermediaries in turning intentions into implementations. They helped
make possible one or more potential implementations of science curricular intentions.

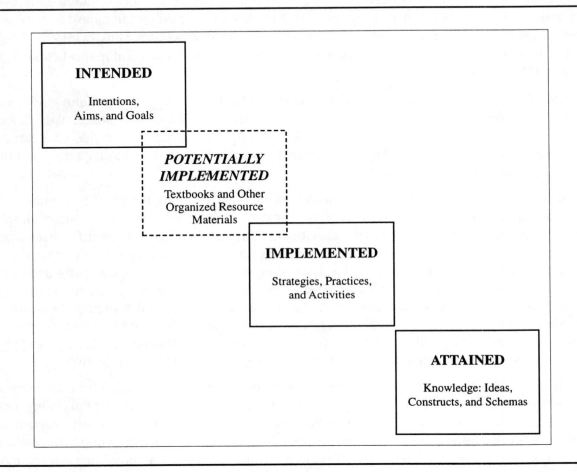

NOTES

1   Jackson, P. W. 1992. "Conceptions of Curriculum and Curriculum Specialists. " In P.W. Jackson, ed., *Handbook of Research on Curriculum.* New York: Macmillan Publishing Co.

2   See, for example: L. C. Comber and J. Keeves, *Science Education in Nineteen Countries* (New York: Wiley, 1973). Also see Marlaine Lockhead, "School and Classroom Effects on Student Learning Gain: The Case of Thailand," paper presented at the annual meeting of the AERA (20-24 April, 1987); J.P. Farrell and S. P. Heyneman (eds.), *Textbooks in the Developing World: Economic and Educational Choices* (New York: The World Bank, 1989) and Brian Holmes and Martin McLean, *The Curriculum: A Comparative Perspective* (New York: Routledge, 1989); Egil Børre Johnsen, *Textbooks in the Kaleidoscope* (Oslo: Scandinavian University Press, 1993).

Appendix B

A MODEL OF POTENTIAL EDUCATIONAL EXPERIENCES

Inferring curricular intentions and aims by analyzing curriculum materials requires assumptions about how intentions affect classroom organization and activities. Such assumptions are the basis for any document analysis methodology. These assumptions and relationships are best portrayed in a model. That is, one needs a thorough model of potential educational experience since that is the key aspect in which curriculum is viewed here.

Curriculum designers make crucial decisions on how students should spend classroom time — choices based on national aims for science education. The set of aims pursued must be limited since each student's time is limited — that is, emphasizing one science topic must often be at the expense of time for another.

These aims and choices affect what happens in classrooms. The system by which they do so is intricate. Other factors, including textbook markets, teacher training and credentialing, school and classroom inspections, examinations and university admission policies, also affect classroom activities. "Official" curricular intentions and aims — as set out in curriculum guides and reflected (at least in part) in textbooks — can help shape classroom activities but not completely determine them. A model of potential experiences must include not only official intentions, but also the goals of other planners and organizers, how intentions are carried out, how intentions and actions interact to produce learning, and other factors. TIMSS used a Model of Potential Educational Experiences that captures some of the important aspects of how educational opportunities are shaped and how they are related to official intentions and aims.

The model (see figure B.1) is built on four broad research questions (the columns) and four educational system levels (the rows). Within the cells created by combinations of these questions and system levels are major constructs that describe educational opportunity — and science learning opportunities in particular. This model and its research questions are described more fully in a related technical report,[1] a part of the TIMSS effort to characterize educational opportunity.

This model involves many constructs, and investigating each typically requires multiple sources of evidence. This volume focuses on part of the model — the characterization of intended student attainments. Even this limited part of the model involves a variety of relevant

Figure B.1
The TIMSS Model of Potential Educational Experiences.
The means by which curricular aims and choices affect what happens in classrooms,
and what students attain is part of an intricate system that includes many factors.

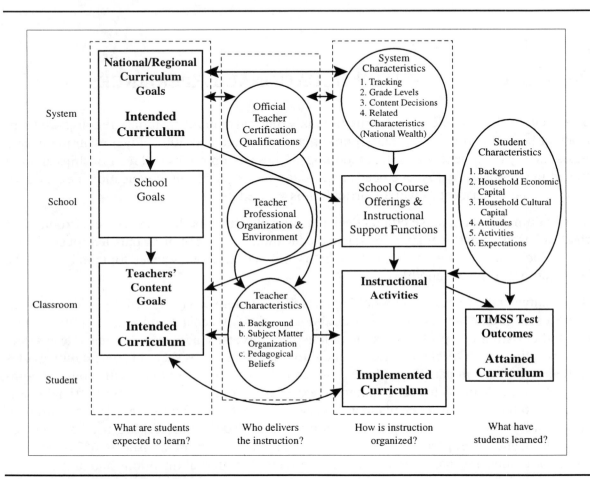

information: ministerial policy documents, formal curriculum guidelines, course syllabi, textbooks, syllabi for national examinations, teacher pedagogical plans as they interpret broader
requirements, tests, and so on. To focus on central documents common to virtually all participating countries, attention here was limited to science curriculum guides and student textbooks.
Supplemental materials and annotated "teacher editions" of textbooks offer additional information but were less consistently used and thus not included in this initial study. While guided by
the model above, the present investigations are limited (1) by focusing on curricular intentions
and (2) by using only curriculum guidelines and textbooks as data sources.

NOTE

[1] Survey of Mathematics and Science Opportunities, *TIMSS Curriculum Analysis: A Content Analytic Approach*, (East Lansing, MI: SMSO, 1993).

Appendix C

TIMSS CURRICULUM FRAMEWORKS: MEASURING CURRICULAR ELEMENTS

The Third International Mathematics and Science Study (TIMSS) was designed to involve extensive data collections from participating national centers, an analysis of curriculum documents, a range of questionnaires, and complex attainment testing. This multi-component data collection (carried out in participating countries and at several international sites) made it essential that all TIMSS components be linked by a common category framework and descriptive language. Whether classifying a test item, characterizing part of a curriculum document, or linking a questionnaire item to other TIMSS parts, all descriptions had to use common terms, categories, and standardized procedures to assign numerical codes for entry into the appropriate database.

This common language was provided by two framework documents: one for the sciences and one for mathematics. Each covered the full range of the years of schooling in a unified category system. Each framework was articulated first only in technical reports (as is still true for more extensive explanatory notes) but has now been reproduced and documented in a monograph.[1]

Each framework document was multi-faceted and multi-layered. It considered three aspects of subject matter content and performance: *content* (subject matter topic), *performance expectation* (what students were expected to do with particular content), and *perspective* (any overarching orientation to the subject matter and its place among the disciplines and in the everyday world). Content as subject matter topic is relatively clear; performance expectation is less so. It was decided not to attempt identifying student (cognitive) performance processes, which would be highly inferential, particularly subject to cultural differences, and thus not feasible in a cross-cultural context. It was decided rather to specify expectations for student (science) performances — for example, formulating or clarifying a problem to be solved, developing a solution strategy, verifying a problem solution, and so on, without positing cognitive means for producing the performances. This focused on more generic, less culture-bound task expectations and demands. Postulating specific cognitive processes necessarily would have involved culturally tied cognitive categories inherent in student thinking in any given country.

In addition to being a multi-aspect system, each framework was designed for using multiple categories. Each element (science curriculum guide, textbook segment, test item) could be considered as part of more than one category of any framework aspect. Each element was to be classified in as many framework categories as needed to capture its richness. Each would have a unique, often complex "signature" — a set of content, performance expectation, and perspective categories that characterized it, at least in terms of the framework's three aspects. This system is flexible. It allows for simple or more complex signatures as needed.

Such flexible multi-dimensionality was essential for analyzing curriculum documents. How often segments involved multiple contents or performance expectations had to be determined empirically by the documents, not by the study design. Characterizing science curricula required a tool permitting coherent categorization of curriculum guides' and textbooks' major pedagogical features. It had to be capable of translating many nationally idiosyncratic ways for specifying science education goals into a common specification language. The TIMSS science curriculum framework was designed to be such a tool.

Framework development was cross-national. The frameworks had to be suitable for all participating countries and educational systems. Representing the interests of many countries, the frameworks were designed cross-nationally and passed through several iterations. The result is imperfect, but it is still a step forward in cross-national comparisons of curricular documents. Its value may be judged by the results of this and related volumes.

Each framework aspect is organized hierarchically using nested sub-categories of increasing specificity. Within a given level, the arrangement of topics does not reflect a particular rational ordering of the content. See figure C.1 for an overview of the content aspect of the science framework. The end of this appendix briefly presents all three aspects of the science framework. Each framework was meant to be encyclopedic — intended to cover all possibilities at some level of specificity. No claim is made that the "grain size" — the level of specificity for each aspect's categories — is the same throughout the framework. Some sub-categories are more inclusive and commonly used, others less so. Specificity had broad commonalties but also considerable variation among the participating countries. This varying granularity requires special care in designing framework-based methods and interpreting the results of their use.

In the science framework, *content* involves eight major categories, each with two to six sub-categories.[2] Some sub-categories are divided further. The level of detail and organization reflects compromise between simplicity (fewer categories) and specificity (more categories). The hierarchical levels of increasing specificity allow some flexibility in detail level and generalization.

Performance expectation is a flexible category system (see below for details), but — as with the other framework aspects — no category or sub-category is considered exclusive. Any document segment should involve at least one or more performance expectation categories from

Figure C.1

Content Categories of the Science Framework.

Each aspect of the framework contained a set of main categories. Each main category contained one or more levels of more specific sub-categories. The main content categories are shown here with some sub-categories expanded to give a better insight into the framework's structure.

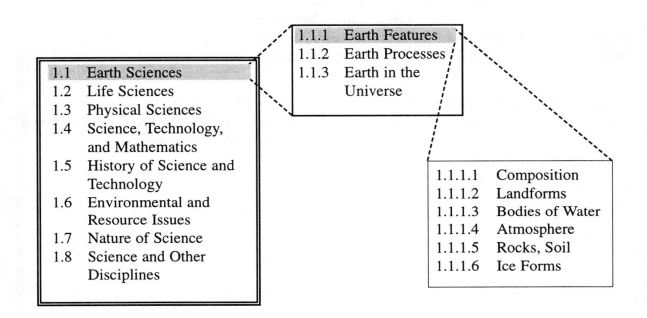

the framework. Complex, integrated performances can thus be characterized in detail — as can contents and perspectives. This differs from traditional grid classifications that generate unique categorizations combining one element from two or more dimensions.

Complex signatures reveal important differences in how curricula are meant to achieve their goals. They show differences in how subject matter elements are combined and differences in what students are expected to do. Each framework can reveal subject matter presented in an integrated, thematic way, with a rich set of performance expectations for students as recommended by curriculum reformers in many countries. However, it also allows simpler "signatures" — for example, those often associated with more traditional curricula and many traditional achievement test items.

The following is a brief listing of the content, performance expectation, and perspective codes of the science framework. For a more detailed discussion, see the monograph referenced or the technical report of "Explanatory Notes" for the science framework.

1.1 Earth Sciences

1.1.1 Earth Features

1.1.1.1 Composition

1.1.1.2 Landforms

1.1.1.3 Bodies of Water

1.1.1.4 Atmosphere

1.1.1.5 Rocks, Soil

1.1.1.6 Ice Forms

1.1.2 Earth Processes

1.1.2.1 Weather and Climate

1.1.2.2 Physical Cycles

1.1.2.3 Building and Breaking

1.1.2.4 Earth's History

1.1.3 Earth in the Universe

1.1.3.1 Earth in the Solar System

1.1.3.2 Planets in the Solar System

1.1.3.3 Beyond the Solar System

1.1.3.4 Evolution of the Universe

1.2 Life Sciences

1.2.1 Diversity, Organization, Structure of Living Things

1.2.1.1 Plants, Fungi

1.2.1.2 Animals

1.2.1.3 Other Organisms

1.2.1.4 Organs, Tissues

1.2.1.5 Cells

1.2.2 Life Processes and Systems Enabling Life Functions

1.2.2.1 Energy Handling

1.2.2.2 Sensing and Responding

1.2.2.3 Biochemical Processes in Cells

1.2.3   Life Spirals, Genetic Continuity, Diversity

1.2.3.1 Life Cycles

1.2.3.2 Reproduction

1.2.3.3 Variation and Inheritance

1.2.3.4 Evolution, Speciation, Diversity

1.2.3.5 Biochemistry of Genetics

1.2.4   Interactions of Living Things

1.2.4.1 Biomes and Ecosystems

1.2.4.2 Habitats and Niches

1.2.4.3 Interdependence of Life

1.2.4.4 Animal Behavior

1.2.5   Human Biology and Health

1.2.5.1 Nutrition

1.2.5.2 Disease

1.3 Physical Sciences

1.3.1   Matter

1.3.1.1 Classification of Matter

1.3.1.2 Physical Properties

1.3.1.3 Chemical Properties

1.3.2   Structure of Matter

1.3.2.1 Atoms, Ions, Molecules

1.3.2.2 Macromolecules, Crystals

1.3.2.3 Subatomic Particles

1.3.3   Energy and Physical Processes

1.3.3.1 Energy Types, Sources, Conversions

1.3.3.2 Heat and Temperature

1.3.3.3 Wave Phenomena

1.3.3.4 Sound and Vibration

1.3.3.5 Light

1.3.3.6 Electricity

1.3.3.7 Magnetism

1.3.4 Physical Transformations

1.3.4.1 Physical Changes

1.3.4.2 Explanations of Physical Changes

1.3.4.3 Kinetic Theory

1.3.4.4 Quantum Theory and Fundamental Particles

1.3.5 Chemical Transformations

1.3.5.1 Chemical Changes

1.3.5.2 Explanations of Chemical Changes

1.3.5.3 Rate of Change and Equilibria

1.3.5.4 Energy and Chemical Change

1.3.5.5 Organic and Biochemical Changes

1.3.5.6 Nuclear Chemistry

1.3.5.7 Electrochemistry

1.3.6 Forces and Motion

1.3.6.1 Types of Forces

1.3.6.2 Time, Space, and Motion

1.3.6.3 Dynamics of Motion

1.3.6.4 Relativity Theory

1.3.6.5 Fluid Behaviour

Performance Expectations

2.1 Understanding

    2.1.1   Simple Information

    2.1.2   Complex Information

    2.1.3   Thematic Information

2.2 Theorizing, Analyzing, and Solving Problems

    2.2.1   Abstracting and Deducing Scientific Principles

    2.2.2   Applying Scientific Principles to Solve Quantitative Problems

    2.2.3   Applying Scientific Principles to Develop Explanations

    2.2.4   Constructing, Interpreting, and Applying Models

    2.2.5   Making Decisions

2.3 Using Tools, Routine Procedures, and Science Processes

    2.3.1   Using Apparatus, Equipment, and Computers

    2.3.2   Conducting Routine Experimental Operations

    2.3.3   Gathering Data

    2.3.4   Organizing and Representing Data

    2.3.5   Interpreting Data

2.4 Investigating the Natural World

    2.4.1   Identifying Questions to Investigate

    2.4.2   Designing Investigations

    2.4.3   Conducting Investigations

    2.4.4   Interpreting Investigational Data

    2.4.5   Formulating Conclusions from Investigational Data

2.5 Communicating

    2.5.1   Accessing and Processing Information

    2.5.2   Sharing Information

Perspectives

3.1 Attitudes Towards Science, Mathematics, and Technology

    3.1.1   Positive Attitudes Toward Science, Mathematics, and Technology

    3.1.2   Skeptical Attitudes Towards use of Science and Technology

3.2 Careers in Science, Mathematics and Technology

    3.2.1   Promoting Careers in Science, Mathematics, and Technology

    3.2.2   Promoting Importance of Science, Mathematics, and Technology in Non-technical Careers

3.3 Participation in Science and Mathematics by Underrepresented Groups

3.4 Science, Mathematics, and Technology to Increase Interest

3.5 Safety in Science Performance

3.6 Scientific Habits of Mind

NOTES

1   Robitaille, D. F. et al., *Curriculum Frameworks for Mathematics and Science*, (Vancouver: Pacific Educational Press, 1993).

2   This volume follows a convention in referring to the categories in the science framework as compared to less formal listing of science topics. Categories from the framework will be listed enclosed in single quotation marks — for example, 'biochemistry of genetics.' Informal topic areas and references to topics that are part but not all of a formal category will not have quotation marks — for example, gene expression rather than 'gene expression.' A similar convention will be followed for discussing categories from the other aspects of the framework.

Appendix D

TOPIC TRACING METHODS

A technique called "topic trace mapping" provided a broad multi-grade map of content coverage. Curriculum experts, using their knowledge of curriculum guides from each country, were asked to report — for each, most detailed science framework content category and for the full set of grade levels within their country— in which grade or grades work on each topic was intended to be introduced, the range of grades during which instruction was intended to take place, and any grades for which the topic was to be a focus (to receive special curricular emphasis).

Figure D.1 shows "tracings" or "maps" of the intended curricular coverage for two science topics in a sample of selected countries to show some of the diversity typical of these data. *Grade level* referred to the sequence of years in school and was adjusted to the student ages and educational policies of the countries involved. That is, the same grade might not represent the same modal student age in two different countries depending on the educational systems involved (although little variation was observed). The symbol "-" indicates that experts reported aspects of the topic were typically part of science instruction at that grade level. The symbol "+" indicates that experts reported that aspects of that topic received special curricular attention and emphasis of some sort at that grade level.

'Biomes and ecosystems' refers to all treatments of these biological topics. The figure shows, for example, that content from 'biomes and ecosystems' received only 2 years of coverage (grades 7 and 8) in Hungary, with focused instruction in grade 7. By contrast, Mexican experts indicated that 8 years of study were intended for this content starting in grade 2 — with special emphasis in grade 7 and then again in grades 11 and 12 (and in both these grades the topic received considerable curricular focus). Other patterns emerged, ranging from countries planning instruction in this topic only in upper secondary (for example, 1 year each in French-speaking Belgium and Tunisia, 4 years in the People's Republic of China in this list of selected countries) to those that intended coverage of the topic from almost the beginning of primary to the end of upper secondary school (Portugal, France, and the Dominican Republic). Other countries show patterns of interrupted or multi-sited coverage, (for example, Latvia, Greece, and Japan).

'Energy types, sources, and conversions' provides a contrast. This topic includes the treatment of potential and kinetic energy, work, efficiency, as well as sources of energy, and so on — in short, all energy-related content in the physical sciences. Experts indicated that it was

Figure D.1
Two Representative Topic Trace Maps.

These data are typical topic trace maps for a sample of countries selected to show representative diversity among maps produced by topic trace mapping. The results are typical of those for other topics and countries.

Biomes & Ecosystems

	\	\	\	\	\	\	Grade	\	\	\	\	\
	1	2	3	4	5	6	7	8	9	10	11	12
Belgium (Fr)											+	
China, People's Republic of								+	-	-	+	
Dominican Republic	-	-	-	-	-	+	-	-	-		-	-
France	-	-	-	-	-	+	-	-		+	-	-
Greece					-	+	+	-	-		-	+
Hungary							+	-				
Israel					+	-	+	-	-	+	+	+
Japan						-			-		+	+
Latvia		-	-	-			-			-	-	
Mexico		-	-	-	-	-	+				+	+
Norway					-	-	-	-		+	-	+
Philippines				-	-	+	-	+	-	-		
Portugal	-	-	-	-	-		+	-	-	-	-	+
Russian Federation				-	-	-	+	+	-	+	-	
Tunisia											+	

Energy Types, Sources, Conversions

	\	\	\	\	\	\	Grade	\	\	\	\	\
	1	2	3	4	5	6	7	8	9	10	11	12
Belgium (Fr)									-	-	+	-
China, People's Republic of								+	+			
Dominican Republic					-	-		-		-	-	-
France					-	-			-	-	+	-
Greece			-	-	+	-	-	+	-	-	-	+
Hungary							+	-	-	-		
Israel	+	-	+	-	-	-	-	-	+	+	+	+
Japan								-	+	-	-	-
Latvia			-						+	+	+	+
Mexico	-	-		-		-	-	+	-	-	+	+
Norway	-	-	-	-		-	-	-	+	-	+	-
Philippines					-	-	+	-	-	+	+	
Portugal				-	-		-	-	+	+	+	
Russian Federation							+	+	+	+	+	
Tunisia					-	-					+	

Legend: - indicates intended coverage at that age.
+ indicates intended focal coverage at that age.

taught primarily in secondary school in many countries — as might be expected for topics in physics. However, some countries report coverage of this topic almost throughout the years of schooling (i.e., Israel, Norway), whereas still others report some work on these topics in elementary school, with successive treatment in upper secondary (France and Tunisia).

The information provided by this method is based on soliciting structured expert opinion in each country and only indirectly on official curriculum guides. Clearly, however, the method provides an essential first step in understanding the context of intended content coverage for those grades targeted for achievement testing by TIMSS. These grades are, essentially, the two grades containing the modal number of 9-year-old students, the two for 13-year-old students, and physics specialists in the terminal year of secondary school. As an instance of the value of these data, curriculum guides in the People's Republic of China and the Russian Federation likely would specify differently the objectives for teaching 'biomes and ecosystems' in grade 8. This reflects that the topic was being introduced in the former in grade 8, while students in the latter were likely to have studied it for several years before grade 8. Similarly, the absence of 'energy types, sources, conversions' for grade 4 in Japan and Latvia — compared, for example, to France and Mexico, where experts indicated instruction would have begun earlier — likely implies a difference in how the topic was treated (if at all) in grade 4 textbooks in those countries.

Data gathered by this topic tracing method allows an initial assessment of the variety and flow of content coverage goals across the grades in participating countries. This method has important drawbacks, although it provides essential information. First, it relies on expert opinion. Second, this method proved in pilot work and field trials to yield no useful information on performance expectations and perspectives. Fortunately, topic tracing is only one component of a set of multiple methods that focus mainly on detailed document analysis of curriculum guides and textbooks.

Appendix E

DOCUMENT ANALYSIS METHODS

Document Sampling

Curriculum document sampling was essential for document-based methods. To collect detailed information for the Third International Mathematics and Science Study (TIMSS) targeted grades — those for student achievement testing — countries gathered data from samples of curriculum guides and textbooks for those grades.

Each national document sample comprised, as appropriate for that country, the following text materials:

- The national science curriculum guide or guides (if any) covering each grade;

- Regional, provincial, state, or cantonal science curriculum guides covering each grade (if needed);

- Official national science textbooks (if any); and

- The most widely used commercial textbooks if "officially" provided books were not used.

Each national sample was required to contain curriculum guides and textbooks sufficient to represent materials for no less than one-half of the students in the targeted grades. These document samples were also to cover major regional, school type, etc., strata in each country.

The resulting sample for the 48 countries (those reporting at least some document-based curriculum data) included 77 science curriculum guides for the upper grade of Population 1, 111 for the upper grade of Population 2, and 62 for the physics specialists in Population 3, for a total of 250 curriculum guides. There were 75 science textbooks for the upper grade of Population 1, 155 for the upper grade of Population 2, and 60 for the physics specialists in Population 3, for a total of 290 sampled science textbooks. These are the data sources for document analyses reported in this volume.

Document Analysis

The document sample was large, and with 48 countries the documents' linguistic variety, was daunting. Methods had to be developed that did not require translating large document segments, although each sampled document (as marked in the analysis) was collected and archived. Central to all document analysis methods was operationally specifying how all document analyses, marking procedures, and reporting procedures would be handled. Training native-language speakers in each country to carry out the procedures to produce coded data forms with some English text was also essential. These data, both coded and full-text English, were entered into a central data bank, cleaned, edited, audited, and verified by the countries contributing the data.

This plan was implemented and is described below. Similar procedures were used to code critical features of both curriculum guides and textbooks. Those features were only some of the possibilities considered for analysis; the selected features were considered salient and the most feasible for this first attempt at large-scale, within-language, cross-national document analysis.

Unit Identification

Most documents were both structurally intricate and long — at least for preparing a statistical abstract with minimal English-language documentation. The first step in each document analysis process was to sub-divide each document into smaller "units of analysis" so that more detailed analyses could be accomplished. For curriculum guides, these units were the smallest functional segments of each document — identified by clear separation rules to segment each document into a set of such units, each classified as one of a small number of types capturing a major function in curriculum guide presentations. Curriculum guides unit types were introductions, units stating policies, units stating educational objectives, units specifying science contents, units dealing with issues in science pedagogy, and "other" units.

Table E.1 shows — for the entire document set, both curriculum guides and textbooks — the proportions of different unit types. More than 50 percent of the science curriculum guide units for the upper grade of Populations 1 and 2, as well as the physics specialists of Population 3, were devoted to specifying objectives and content. Science curriculum guides for the upper grades of both Populations 1 and 2 had an average of 17 to 19 other units. Further, policy units were not very common on average — representing 5 percent or less of the units.

Textbooks were similarly partitioned. The most fundamental unit type was a *lesson*. A "lesson" was a text material segment devoted to a single main science topic and intended to correspond to a teacher's classroom lesson on that topic taught over one to three instructional periods. Some textbook materials (for example, reviews) covered content from several lessons and were classified as "multiple-lesson pages." These two unit types, along with introductions, instructional appendices, and "other" made up the five textbook unit types.

Each country's data collection team identified and assigned unit types to each textbook unit. Table E.1 shows about 78 percent of the units identified were lessons in the pool of science textbooks sampled. Typically there were less units in textbooks intended for the upper grade of Population 1 (an average of 30) than for the upper grade of Population 2 or for physics specialists in Population 3 (averaging 45 and 68 respectively).

Table E.1

Unit Types in Curriculum Guides and Textbooks.

Percentages of each unit type for the documents sampled are shown for all populations and for both guides and textbooks. Curriculum guides were dominated by objective and content units. Lesson units clearly dominated textbooks.

Curriculum Guides	Introduction to Guide	Policy Units	Objective Units	Content Units	Pedagogy Units	Other Units	Missing
Population 1 Upper Grade	2	4	27	31	13	21	3
Population 2 Upper Grade	4	5	23	37	12	17	3
Population 3 Specialists	2	4	25	38	8	20	3

Textbooks	Introduction	Lesson	Multiple-Lesson Pages	Instructional Appendix	Other	Missing
Population 1 Upper Grade	1	86	8	2	2	2
Population 2 Upper Grade	2	78	15	3	2	1
Population 3 Specialists	1	72	19	3	2	2

Block Identification

Dividing documents into broad functional units simplified analysis; but this procedure was not enough to capture essential structural aspects allowing construction of a "statistical" description of the textbook or at least some of its key features. To capture more features showing differences among science curriculum guides and science textbooks, these texts were divided into smaller segments called 'blocks.'

Figure E.1 is a reproduction of two sample pages from a Colombian science textbook for the upper grade of Population 2. These two pages are the first of Unit 35, a lesson (unit type 2 — "T. 2" here). This unit is on genetics and the dairy industry, part of the treatment of genetic structure and systems in eighth grade science. Following the standardized partitioning rules, these two pages are segmented into seven blocks — each with marked boundaries and numbered 1 through 7.

In a similar way each unit of every sampled science textbook was sub-divided into smaller segments — i.e., 'blocks.' Identifying several different block types for textbooks and others for curriculum guides helped construct a more detailed "statistical abstract" of each document. For curriculum guides, the types included individual policies, single objectives, single science content elements, individual pedagogical suggestions, individual examples related to pedagogical suggestions, individual testing or "assessment" suggestions, and "other" block types.

For science textbooks, block types included narrative blocks (central, related, and unrelated sidebar instructional narrative), graphic blocks related to narrative, graphic blocks unrelated to instructional narrative, exercise and question sets (both directly related to a unit's content and "unrelated," for example, review sets inserted as part of a unit on a different topic), suggested activities, worked examples, and "other" textbook block types. Precise rules available in TIMSS technical reports governed partitioning units into blocks and identifying block types.

As an example of the process, consider figure E.1 again. Even blocks are marked and numbered on the two pages — note that Block 3 is related to Block 1. Block 1 is a brief segment of narrative. This passage explains how the dairy industry uses microorganisms to produce a variety of products and that their specific genetic characteristics produce products of varying aromas, flavors, and consistencies. The specific characteristics of various species of microorganisms and their most common use is presented in the related table (Block 3). This is followed by an activity, to be carried out in class using *Lactobacillus bulgaricus* to produce yogurt. This activity includes producing the yogurt and then testing its viscosity, appearance, etc.— and recording the results of the tests in a table to be devised by the student. In Block 4, students are asked to describe the relationships among a series of scientific concepts, such as "gene, chromosome, DNA" and "ribosomes, messenger RNA, transcriber RNA, DNA." The following Block (5) is a small exercise asking students to define, in their own words, various terms such as "locus," "alleles," and "gametogenesis." The last blocks on page 99 describe a fictional legal

Figure E.1

Two Sample Textbook Pages.

Here are two sample pages from a Columbian science textbook for the upper grade of Population 2. They represent part of one unit. Their division into blocks and the numbers for each block and unit were indicated by the Colombian National Document Analysis Team.

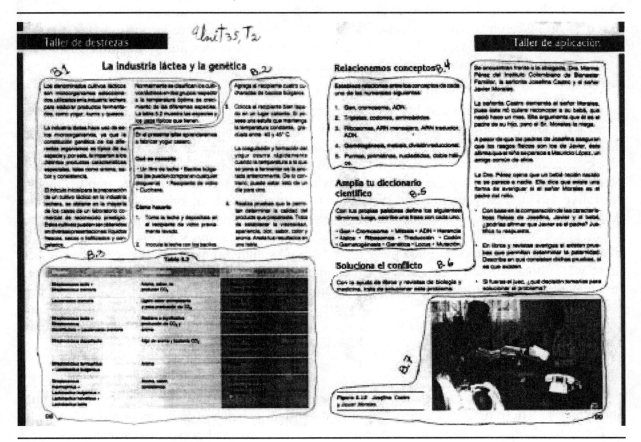

case in which the paternity of a child is in question. Students are asked to research what types of scientific tests are available to "solve the conflict" and to describe their use.

The strategy in the opening pages of this unit is clear. A general narrative introduces the topic of applications of genetics. An activity is presented. A review exercise and a dictionary exercise follow. Finally, another application of genetic science is discussed and children are asked to do research on the application. All of this detail must be reduced to a few essential elements that allow the search for commonalities among countries and populations. One element of those details is block type, recording here that the strategy on these two pages was narrative, activities, and exercises.

Table E.2 shows that most blocks in curriculum guides were related to content elements, objectives, and pedagogical suggestions. However, there were fewer pedagogical suggestion blocks in science than was the case in mathematics curriculum guides. Comparatively few blocks contained official policies or assessment suggestions. In textbooks, roughly 33 percent of all blocks — in all of the grades from which textbooks came — were graphic blocks. Narrative (written explanation and discussion) — averaged about one-quarter of the blocks. Activity blocks diminished for each successive population, going from about 15 percent for the upper grade of Population 1, to scarcely 3 percent in Population 3 physics texts. However, overall, science textbooks still had more activity blocks than did mathematics textbooks, where there were more worked examples. Perhaps the most striking contrasts between science and mathematics textbooks was in the relative occurrence of exercises and graphics. Almost one-half the blocks in mathematics textbooks were devoted to exercises, while only about 15 percent of the blocks in science texts were. Thirty percent of science textbook blocks were graphics, whereas only about 10 percent was the norm in mathematics.

Coding

Central to preparing "abstracts" of documents was partitioning and assigning types (units and blocks). These data and their relation to document pages had to be recorded. It was also essential to identify the contents, performance expectations, and perspectives relevant to each block. After dividing a document into units and blocks, each block within each unit was described by assigning to it codes for categories from each aspect of the science framework. These codes — along with block and unit types — were the main information recorded for the "statistical abstract" characterizing each document. Some other descriptive information was also provided as described in appendix A.

Figure E.2 shows how members of a national data collection team record these categories on paper forms with columns corresponding to blocks within each unit. This reproduction of the coding sheet used by national coders to record block numbers, links to document pages, block numbers, block types, and science framework codes was completed by a reviewer of the Colombian text segment reproduced in figure E.1. The first seven columns are for the seven

Table E.2
Distribution of Block Types.

Here are the percentages of block types for the pool of sampled curriculum guides and textbooks. Curriculum guides were mainly composed of objective, content, and pedagogical suggestion blocks. Textbooks mainly contained related graphic and narrative blocks. Activities played a larger role for younger populations; exercise sets played a larger role for older populations.

Curriculum Guides	Official Policies	Objectives	Content Element	Pedagogical Suggestions	Examples	Assessment Suggestions	Other	Missing
Population 1 Upper Grade	6	24	36	16	11	3	3	1
Population 2 Upper Grade	7	26	39	14	9	3	2	1
Population 3 Specialists	6	26	40	11	12	3	2	1

Textbooks	Narrative	Related Narrative	Unrelated Instructional Narrative	Related Graphic	Unrelated Graphic	Exercise/ Question Set
Population 1 Upper Grade	27	4	1	35	2	15
Population 2 Upper Grade	30	8	1	31	2	17
Population 3 Specialists	28	9	2	34	1	17

	Unrelated Question/ Exercise Set	Activity	Worked Example	Other	Missing
Population 1 Upper Grade	1	16	0	1	0
Population 2 Upper Grade	1	8	1	1	1
Population 3 Specialists	0	3	6	1	0

Figure E.2

Sample Block Coding Form.

This is the completed coding form that includes the blocks from the earlier sample text pages (Figure E.1). It records a block's page number (in the text), identification number, type, and codes for appropriate framework categories.

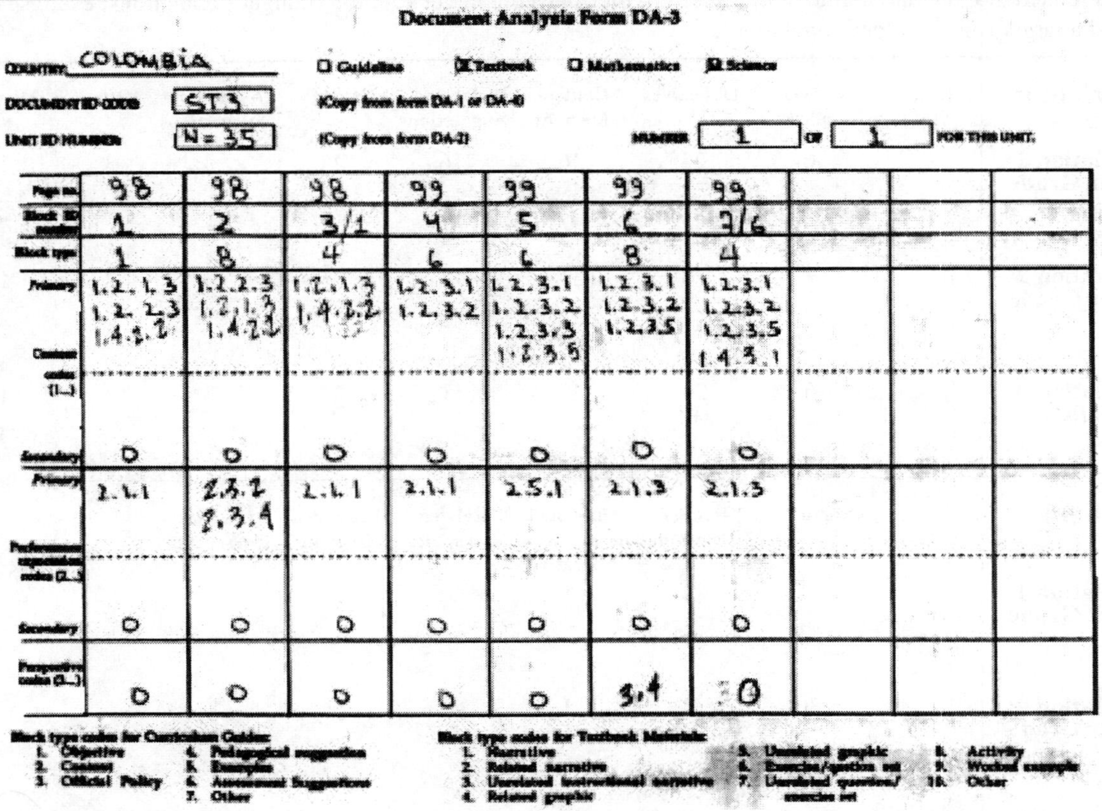

Document Analysis Form DA-3

COUNTRY: COLOMBIA ☐ Guidelines ☒ Textbook ☐ Mathematics ☒ Science

DOCUMENT ID CODE: ST3 (Copy from form DA-1 or DA-0)

UNIT ID NUMBER: N=35 (Copy from form DA-2) NUMBER 1 OF 1 FOR THIS UNIT.

Page no.	98	98	98	99	99	99	99			
Block ID number	1	2	3/1	4	5	6	3/6			
Block type	1	8	4	6	6	8	4			
Content codes (1...) Primary	1.2.1.3 1.2.2.3 1.4.1.2	1.2.2.3 1.2.1.3 1.4.2.2	1.2.1.3 1.4.3.2	1.2.3.1 1.2.3.2	1.2.3.1 1.2.3.2 1.2.3.3 1.2.3.5	1.2.3.1 1.2.3.2 1.2.3.5	1.2.3.1 1.2.3.2 1.2.3.5 1.4.3.1			
Content codes Secondary	0	0	0	0	0	0	0			
Performance expectation codes (2...) Primary	2.1.1	2.3.2 2.3.4	2.1.1	2.1.1	2.5.1	2.1.3	2.1.3			
Performance expectation codes Secondary	0	0	0	0	0	0	0			
Perspective codes (3...)	0	0	0	0	0	3.4	0			

Block type codes for Curriculum Guides:
1. Objective
2. Content
3. Official Policy
4. Pedagogical suggestion
5. Examples
6. Assessment Suggestions
7. Other

Block type codes for Textbook Materials:
1. Narrative
2. Related narrative
3. Unrelated instructional narrative
4. Related graphic
5. Unrelated graphic
6. Exercise/question set
7. Unrelated question/exercise set
8. Activity
9. Worked example
10. Other

blocks in the text segment. Notice the block types are recorded by numerical code (1, 8, 4, and 6) — and linked both to textbook page and block number. The first portion of text on page 98 has been assigned a block type of 1, which indicates narrative. It has also been assigned three content codes: 1.2.1.3 refers to types of microorganisms; 1.2.2.3 is the science category framework for 'biochemical processes in cells'; and the code 1.4.2.2 indicates industrial applications. The performance expectation code assigned is 2.1.1, which indicates that this portion of the text merely requires students to understand 'simple information.' Block 3 is the table associated with Block 1, which is indicated by the notation "3/1" in the "block ID number" cell. The block type is 4, indicating a related graphic; it is also assigned the content codes of 1.2.1.3 and 1.4.2.2.

Block 2 is assigned a type of 8, indicating an activity. The content codes assigned are identical to those for Block 1; however, the performance expectation codes are quite different. 2.3.2 indicates 'conducting routine experimental operations,' and 2.3.4 refers to 'organizing and representing data' (part of the activity is to devise a table to record data). Blocks 4 and 5 are the two exercises (type 6), one assigned a performance expectation code of 2.1.1, the other a new code, 2.5.1 for 'accessing and processing information,' indicating that students are expected to use a reference work (dictionary). Finally, Blocks 6 and 7, an activity (type 8) and its related graphic (type 4) refer to the material on determining paternity in the context of a legal dispute. Here, in addition to the content codes used in previous blocks, the coder has assigned 1.4.3.1, the framework category for 'influence of science, technology, on society.' Block 6 is also assigned a perspective code of 3.4, the framework code indicating 'science, mathematics, and technology to increase interest,' signifies that popular or intriguing information is used to increase student interest in the topic of genetics. The rare use of perspective codes was a typical pattern. Notice also that all blocks have complex "signatures" — all blocks on the form received more than one code for a single aspect from the framework.

The complexity of block signatures — the number of categories from each aspect of the framework assigned to a block — is important. Complicated signatures reveal complexity and suggest higher levels of integration in curriculum guides and textbooks. Table E.3 shows the percentage of blocks that received single and multiple content codes in science curriculum guides and textbooks at each of the three TIMSS populations. For content, for all three target grades, around 43 percent of the curriculum guide blocks were assigned one code and 20 percent assigned two. In textbooks, material is more complex, with a quarter of the blocks receiving at least two content codes. Overall, science shows an increase in complex signatures for higher grades in curriculum guides — and this is somewhat true for textbooks as well.

Table E.4 summarizes how science curriculum guides and textbooks differed in the number of performance expectation codes assigned. About 25 percent of the blocks in science curriculum guides had no performance expectations indicated. This proportion was far lower for textbooks. About one-half of the blocks in science curriculum guides had one performance expectation code assigned, as did about 85 percent of textbooks. The number of blocks assigned more than one performance expectation was much greater in curriculum guides (about 30

Table E.3
Proportion of Single and Multiple Content Codes.
The average number of content codes assigned to a block is given for each type of document and population. The percentages of single and multiple codes among all those assigned to blocks are also given. Clearly, single and double codes predominated.

Curriculum Guides	Average Number of Content Codes/Block	Distribution of Number of Content Codes					
		0	1	2	3	4	5 or more
Population 1 **Upper Grade**	1.14	30	40	20	7	2	1
Population 2 **Upper Grade**	1.28	23	43	22	9	2	1
Population 3 **Specialists**	1.25	22	45	23	7	2	1
Textbooks							
Population 1 **Upper Grade**	1.38	4	63	25	7	1	0
Population 2 **Upper Grade**	1.45	3	62	25	8	2	0
Population 3 **Specialists**	1.49	1	64	25	8	2	0

Table E.4

Proportion of Single and Multiple Performance Expectation Codes.

The average number of performance expectation codes assigned to a block is given for each type of document and population. The percentages of single and multiple codes are also given.

Curriculum Guides	Average Number of Content Codes/Block	Distribution of Number of Content Codes					
		0	1	2	3	4	5 or more
Population 1 Upper Grade	1.10	28	44	20	7	1	1
Population 2 Upper Grade	1.12	25	49	20	5	1	1
Population 3 Specialists	1.22	22	45	23	8	1	1
Textbooks							
Population 1 Upper Grade	1.06	8	81	8	2	0	0
Population 2 Upper Grade	1.06	5	87	6	2	0	0
Population 3 Specialists	1.05	3	91	5	1	0	0

percent of blocks) than textbooks (merely 8 percent). This would appear to indicate that text-books are more conservative in performance expectations for students than are curriculum guides. Thus, while simpler signatures predominated in both, there was a significant minority of blocks in curriculum guides with more complex performance demands. The case for per-spective codes differed, with 83 percent or more of blocks in science curriculum guides assigned no perspective codes and more than 90 percent of the blocks in textbooks assigned none.

Training, Reliability and Quality Assurance

Data collection methods were demanding, as can be seen even from this brief sketch. More details can be found in TIMSS technical reports. This demanding data collection required addition-al procedures to ensure high data quality, reliability, and validity. The approaches used were face-to-face training, an initial assessment of reliable blocking and coding (with retraining if necessary), careful monitoring and country review of all data submitted, and assessment of coding reliability.

Representatives responsible for science and mathematics curriculum data collection in every TIMSS country attended regional training on the procedures outlined above. Training used a standard 3-day design with training of trainers, complete manuals documenting all aspects of the standardized training and all procedures, and initial consultations on each coun-try's documents. These training sessions used standardized presentations, training-to-criterion exercises, and more than 400 printed pages of training materials to ensure a thorough, relative-ly uniform initial understanding of the data collection techniques. Representatives of 50 coun-tries attended these sessions, including all those eventually taking part in the curriculum analy-sis. Each "trainee" was requested to conduct similar training sessions in his or her own coun-try for all personnel used to collect and document data.

An initial quality assurance phase was used to verify the adequacy of procedure use imme-diately after training. Those responsible for each country's curriculum data were asked to sub-mit data samples collected from each coder to the international center coordinating curriculum data collection before proceeding with the rest of data collection. These samples were evaluat-ed by comparing independent analyses of translated versions by international referees. National curriculum analysis coordinators were provided with detailed comments on each coder's work. Based on this evaluation, countries were requested either to proceed with data collection or to train their personnel and/or to provide additional work samples. Only when each country coder had demonstrated criterion concordance with international referees were countries asked to continue with the collection of final data.

Final data were examined as they arrived at the international center coordinating the cur-riculum analysis. All questions of accuracy and conformity with procedures were identified and clarified in communication with the data collection teams in each country. Cleaned and clari-fied data were entered into a central database and subjected to further error checks and assessment of data entry accuracy. Cleaned data were subjected to a reliability analysis. Since there were no complete translations of all documents, a traditional inter-rater reliability estimate was clearly inappropriate. A more complex procedure was developed (and is being described in a technical report currently in preparation). Essentially, a random sample of units from the database was translated (if not in English), recoded by referees, and compared to the country codings of the same material. These comparisons showed concordance indices over 80 percent in virtually all cases and often 90 percent or more.

Appendix F

PARTICIPATING COUNTRIES

Argentina	ARG	Japan	JPN
Australia	AUS	Korea	KOR
Austria	OST	Latvia	LVA
Belgium (Flemish-speaking)	BFL	Lithuania	LTU
Belgium (French-speaking)	BFR	Mexico	MEX
Bulgaria	BGR	Netherlands	NLD
Canada	CND	New Zealand	NZL
China, People's Republic of	CHN	Norway	NOR
Colombia	COL	Philippines	PHL
Cyprus	CYP	Portugal	PRT
Czech Republic	CZE	Romania	ROM
Denmark	DNK	Russian Federation	RUS
Dominican Republic	DRP	Scotland	SCO
France	FRA	Singapore	SGP
Germany	DEU	Slovak Republic	SVK
Greece	GRC	Slovenia	SVN
Hong Kong	HKG	South Africa	SAF
Hungary	HNG	Spain	ESP
Iceland	ISL	Sweden	SWE
Iran	IRN	Switzerland	SUI
Ireland	IRL	Thailand	THA
Israel	ISR	Tunisia	TUN
Italy	ITA	United States of America	USA

ARGENTINA
Prof. Carlos Mansilla
Universidad del Chaco

AUSTRALIA
Dr. Jan Lokan
Australian Council for
Educational Research (ACER)

AUSTRIA
Dr. Günter Haider
Universität Salzburg
Institut für Erziehundswissenschaft

BELGIUM (Flemish-speaking)
Dr. Christiane Brusselmans-Dehairs
Rijksuniversiteit Ghent
Seminarie en Laboratorium voor Didactiek

(French-speaking)
Prof. Georges Henry
SEDEP, Université de Liège

BULGARIA
Dr. Kiril G. Bankov
Foundation for Research,
Communication, Education and Informatics
(INCOBRA)

CANADA
Dr. Alan Taylor
University of British Columbia

Mr. Jim Brackenberry
Ministry of Education, Alberta

Allan Craig
Ministry of Education and Training, Ontario

CHINA
Prof. Meng Hong Wei
The China National Institute of
Educational Research

COLOMBIA
Dr. Carlos Jairo Diaz
Universidad del Valle

CYPRUS
Dr. Constantinos Papanastasiou
University of Cyprus

CZECH REPUBLIC
Mr. Vladislav Tomásek
Research Institute of Education

Dr. Jana Straková
Research Institute of Education

DENMARK
Dr. Børge Prien
The Danish National Institute for
Educational Research

DOMINICAN REPUBLIC
Prof. Sarah Gonzalez
Pontificia Universidad Católica Madre y
Maestra

Dr. Eduardo Luna
Barry University

FRANCE
Dr. Emilie Barrier
Centre International d'Études
Pédagogiques (CIEP)

Dr. Anne Servant
Ministère de l'Éducation Nationale

GERMANY
Dr. Rainer H. Lehmann
Universität Hamburg

GREECE
Dr. Georgia Kontongiannopoulo-Polydorides
University of Athens

HONG KONG
Dr. Frederick Leung
University of Hong Kong

HUNGARY
Dr. Péter Vári
National Institute of Public Education

ICELAND
Dr. Einar Gudmundson
Ministry of Education and Culture

IRAN
Dr. Ali Reza Kiamanesh
Ministry of Education

IRELAND
Ms. Deirdre Stuart
St. Patrick's College
Educational Research Centre

ISRAEL
Dr. Pinchas Tamir
The Hebrew University

ITALY
Dr. Anna Maria Caputo
Ministerio della Pubblica Instruzione

JAPAN
Dr. Masao Miyake
Mr. Yuri Saruta
National Institute for Educational Research

KOREA
Dr. JinGyu Kim
National Board of Educational Evaluation

LATVIA
Dr. Andrejs Geske
University of Latvia

LITHUANIA
Dr. Algirdas Zabulionis
University of Vilnius

MEXICO
Dr. Fernando Córdova Calderón
Secretaría de Educación Pública
Evaluación de Políticas y Sistemas Educativos

NETHERLANDS
Dr. Wilmad Kuiper
University of Twente

NEW ZEALAND
Dr. Hans Wagemaker
Ministry of Education

NORWAY
Dr. Svein Lie
University of Oslo

PHILIPPINES
Dr. Milagros D Ibe
University of the Philippines

Dr. Ester B. Ogena
Science Education Institute, Department of
Science and Technology

PORTUGAL
Dr. Gertrudes Amaro
Instituto de Inovaçao Educacional

ROMANIA
Ms. Gabriela Noveanu
Institute of Educational Sciences

RUSSIAN FEDERATION
Dr. Galina S. Kovalyova
The Russian Academy of Education

SCOTLAND
Mr. Brian Semple
Scottish Office Education Department

SINGAPORE
Ms. Mok Siew Eng
Ministry of Education

SLOVAK REPUBIC
Dr. Maria Berova
Vyskummy ústav pedagogicky

SLOVENIA
Mr. Marjan Setinc
Pedagoski Institut, Ljubljana

SOUTH AFRICA
Dr. Derek Gray
Human Sciences Research Council

SPAIN
Mr. José Antonio
López Varona
Instituto de Calidad y Evaluación (INCE)

SWEDEN
Dr. Kjell Gisselberg
University of Umeå

SWITZERLAND
Mr. Erich Ramseier
Amt für Bildungsforschung der
Erziehungsdirektion des Kantons Bern

THAILAND
Dr. Suwaporn Semheng
Institute for the Promotion of Teaching
Science and Technology (IPST)

TUNISIA
Dr. Mahdi Abdeljaouad
Institut Supérieur de l'Education
de la Formation Continue

USA
Dr. William H. Schmidt
Michigan State University

Appendix G

DOCUMENTS ANALYZED

Country/Document Type	Document Title/Publisher/Date

Argentina

Textbook

Biologia 1. Kapelusz, 1985
Fisica 2. Kapelusz, 1992
Serie Dinamica-Ciencias Naturales. Kapelusz, 1987

Australia

Curriculum Guide

Physics Study Design. Victoria Curriculum and Assessment Board, 1991
Science and Technology. K-6. Board of Studies, New South Wales, 1991
The Science Framework P-10. Ministry of Education (Schools Commission), Victoria.
Science Syllabus (1984). Secondary School Board, New South Wales, Amended May 1989
Two Unit Course- Physics. Board of Senior School Studies, New South Wales, Department of Education

Textbook

Exploring Science 2. Macmillan Company of Australia, 1986
Exploring Together Book 2. Martin Educational Horwitz Grahame Pty Ltd., 1988
Fundamentals of Senior Physics. Heinemann, 1985
Look! Book 2. Australian Science and Technology, Longman Cheshire, 1991
Physics Two: Sound, Electronics, Electric Power Motion, Gravity, Structures, Light and Matter. Rigby Heinemann (Octopus Publishing Group Australia Pty Ltd), 1991
Secondary Science 2. John Wilkinson, MacMillan Co. of Australia Pty Ltd., 1983

Austria

Curriculum Guide	*Lehrplan der AHS-Vollstandige Ausgabe 3.* Österreichischer Bundesverlag, 1989
	Lehrplan der Hauptschule. Österreichischer Bundesverlag, 1989
	Lehrplan der Volksschule. Österreichischer Bundesverlag
Textbook	*Biology und Umweltkunde - 8.* Veritas, 1991
	Physik und Chemie 3 (Part A and B). Ueberreuter, 1987
	Physik und Chemie 4 (Part A and B). Ueberreuter, 1990
	Sachunterricht 4. Österreichischer Bundesverlag, 1992
	Sachunterricht 4 - Salzburg. Österreichischer Bundesverlag, 1992

Belgium (Fl)

Curriculum Guide

Leerplan aardrijkskunde voor de eerste graad, tweede gemeenschappelijk jaar - D/1990/4244/1. Gemeenschapsonderwijs.

Leerplan aardrijkskunde voor de eerste graad, tweede leerjaar. Nationaal Verbond van het Katholiek Onderwijs, 1990

Leerplan aardrijkskunde voor het basisonderwijs. Nationaal Verbond van het Katholiek Onderwijs, 1970

Leerplan aardrijkskunde voor het rijkslager onderwijs. Ministerie van Nationale Opvoeding en Cultuur, 1970

Leerplan biologie, tweede leerjaar van het secundair onderwijs. Nationaal Verbond van het Katholiek Onderwijs, 1988

Leerplan biologie voor de eerste graad, tweede leerjaar - D/1989/4244/22. Gemeenschapsonderwijs

Leerplan biologie, wetenschappelijk werk voor de eerste graad, tweede leerjaar D/1989/4244/22. Gemeenschapsonderwijs

Leerplan fysica, moderne wetenschappen, wetenschappelijk werk - D/1989/0279/050. Nationaal Verbond van het Katholiek Onderwijs.

Leerplan fysica, voor de derde graad ASO - D/1992/0279/014. Nationaal Verbond van het Katholiek Onderwijs.

Leerplan fysica, voor de derde graad, 6de leerjaar doorstroming - D/1980/1984/39. Gemeenschapsonderwijs

Leerplan fysica, voor de eerste graad, tweede leerjaar - D/1980/1984/39. Ministerie van Nationale Opvoeding en Nederlandse Cultuur.

Leerplan fysica, wetenschappelijk werk, eerste graad, tweede leerjaar - D/1980/1984/39. Ministerie van Nationale Opvoeding en Nederlandse Cultuur.

Natuurwetenschappen voor het basisonderwijs. Nationaal Verbond van het Katholiek Onderwijs, 1980

Belgium (Fr)

Curriculum Guide

Construire son milieu. Commentaire du programmes des études. Conseil central de l'Enseignement Primaire Catholique

Enseignement secondaire de type I. Programmes expérimentaux pour les trois degrés. Sciences. MCAP, s.c., 1979

Enseignement secondaire rénové. 3e degré - 6e année, de l'enseignement général: programme de physique. Ministère de l'Éducation Nationale et de la Culture Française, 1975

Étude du milieu: (a) naturel et humain; (b) scientifique et technologique. Ministère de l'Éducation Nationale et de la Culture Française, 1982

Plan d'études et instructions pédagogiques pour les trois degrés des écoles primaires. Ministère de l'Instruction Publique, 1985

Bulgaria

Textbook

Astronomy 3. Prosveta, 1991

Biology 2. Prosveta, 1986

Chemistry 2. Prosveta, 1991

Geometry 2. Prosveta, 1991

Physics 2. Prosveta, 1991

Science 1. Prosveta, 1991

Canada

Curriculum Guide

The Common Curriculum. Ontario Ministry of Education, 1993

Curriculum Guidelines Science Intermediate and Senior Divisions, Part 14. Ontario Ministry of Education, 1988

Elementary Science Curriculum Guide. Ministry of Education (BC), Program & Curriculum Development Branch, 1981

The Formative Years. Ontario Ministry of Education, 1975

Junior Secondary Science Curriculum Guide and Resource Book. Ministry of Education (BC)

Junior High Science Program of Studies. Alberta Education, 1990

Physics 10-20-30 Curriculum Guide. Alberta Education, 1977

Programme d'études primaire - Sciences de la nature. Direction générale des programmes, Ministère de l'éducation du Québec, 1980

Programme d'études secondaire - Sciences physiques. Direction générale des programmes, Ministère de l'éducation du Québec, 1987

Science Intermediate and Senior Divisions, Part 2. Ontario Ministry of Education, 1987

Science is Happening Here. Ontario Ministry of Education, 1988

Textbook   *A la découverte des sciences de la nature.* LIDEC Inc., 1989
                            Focus on Science 4. D.C. Heath Canada Ltd., 1983
                            Focus on Science 8. D.C. Heath Canada Ltd., 1989
                            Fundamentals of Physics: A Senior Course. D.C. Heath Canada
                                Ltd., 1986
                            Initiation à l'étude scientifique de l'environnement. Guérin,
                                1988
                            Physics (2nd Edition). Prentice Hall Inc., 1985
                            Physics- Principles and Problems. Charles E. Merrill Publishing
                                Co., 1986
                            Probe 8. John Wiley & Sons, 1985
                            Science. Addison - Wesley, 1980
                            Science Directions 8. John Wiley & Sons, 1991

China, People's Republic of

Textbook   *The First and Second Volumes of Geography for Junior
                                Secondary Schools.* People's Education Press, 1992
                            *The First and Second Volumes of Physics for Junior Secondary
                                Schools.* People's Education Press, 1991
                            The First Volume of Science For Primary School. People's
                                Education Press, 1989
                            Zoology for Junior Secondary School. People's Education
                                Press,1988

Colombia

Curriculum Guide   *Ciencias Naturales y Salud.* Ministerio de Educación Nacional,
                                1987
                            Ciencias Naturales y Salud. Ministerio de Educación Nacional,
                                1990
                            Programa General de Física. Ministerio de Educación Nacional,
                                1990

Textbook   *Ciencias Naturales 4.* Editorial Santillana, 1989
                            Ciencias Naturales y Salud 8. Editorial Sanistllana, 1991
                            Descubrir 8. Editorial Norma, 1992
                            Física, 11°. Educar Editores, 1991
                            Física Fundamental 2. Editorial Norma, 1993
                            Viva la Ciencia 4. Editorial Norma, 1992

Cyprus

Textbook   *Anthropology and Health Education for the Second Grade of
                                Gymnasium.* Ministry of Education, 1992
                            Biology Lessons. Ministry of Education, 1991
                            Chemistry for the Eight Grade. Ministry of Education, 1991

Physics for Lyceum, 12th Grade, Combination 2. Ministry of Education, 1992

Science for the Fourth Grade of Primary School. Ministry of Education, 1987

Science, Fifth Grade. Ministry of Education, 1992

Science for 6th Grade. Ministry of Education, 1992

Steps in Physics. Ministry of Education, 1992

Steps in Physics, 8th Grade. Ministry of Education, 1992

Steps in Physics for 9th Grade. Ministry of Education, 1992

Czech Republic

Curriculum Guide

Curriculum Guide for Basic School: Elementary Geography Grade 4. Fortuna, 1991

Curriculum Guide of Basic School: Science, Grade 4. Fortuna, 1991

Curriculum Guide of Basic School: Physics, Grades 6 to 8. Fortuna, 1991

Curriculum Guide of Basic School: Life Science and Geology, Grades 5-8. Fortuna, 1991

Curriculum Guide of Basic School: Geography, Grades 5-8. Fortuna, 1991

Curriculum Guide for Gymnasium: Physics. Fortuna, 1991

Curriculum Guide for Basic School: Chemistry, Grades 7 to 8. Roenik, 1991

Textbook

Chemistry for Grade 8 of Basic School. Státní Pedagogické Nakladatelství, 1991

Elementary Geography and History for Grade 4 of Basic School. Fortuna, 1992

Life and Earth Science for Grade 8 of Basic School. Státní Pedagogické Nakladatelství, 1988

Physics for Grade 4 of Gymnasium. Státní Pedagogické Nakladatelství, 1987

Physics for Grade 8 of Basic School. Státní Pedagogické Nakladatelství, 1985

Science for Grade 4 of Basic School. Státní Pedagogické Nakladatelství, 1987

Denmark

Curriculum Guide

Fysik/kemi 1989/2 37.13 Undervisningsvejledning for Folkeskolen. Undervisningsministeriet

Fysik, Bekendtgørelse og vejledende retningslinier. Undervisningsministeriet Direktoratet for Gymnasieskolerne og HF, 1988

Textbook   *Biologi 3. klassetrin. Grundbog.* Gjellerups biologi system, 1982
                    EL-7. DLH Fysisk Institut, Sjælsøskolen Birkerød, 1980
                    Fysikkens spor. Nordisk Forlag A.S., 1991
                    Luft og vand, Ny Fysik/kemi 3. Gyldendal, 1991

Dominican Republic

Curriculum Guide   *Ciencias - Educación Básica - Cuarto Grado.* Secretaría de
                    Estado de Educación, Bellas Artes y Cultos, 1987
                    *La Naturaleza y sus Manifestaciones - Educación Básica -
                    Octavo Grado.* Secretaría de Estado de Educación, Bellas
                    Artes y Cultos, 1987

Textbook   *Estudios Sociales y Ciencias Naturales, Libro 8vo. Grado.*
                    Secretaría de Estado de Educación, Bellas Artes y Cultos, 1993
                    Matematica y Ciencia 4°. Secretaría de Estado de Educación,
                    Bellas Artes y Cultos, 1993

France

Curriculum Guide   *Collèges: Programmes et instructions.* Ministère de l'Education
                    Nationale, de la Jeunesse et des Sports, et CNDP, Hachette,
                    1985 (Sciences et techniques biologiques et géologiques: p.
                    269-271 and 275-277.)
                    Compléments aux programmes de biologie et géologie.
                    Supplement au Bulletin Officiel n°25, 1988.
                    Les Cycles à l'école primaire. Ministère de l'Education
                    Nationale, de la Jeunesse et des Sports, et CNDP, Hachette,
                    1991 (Science et technologie: p 66-68 and p. 109-112.)
                    *Principes directeurs de l'enseignement de la physique et de la
                    chimie au collège et au lycée.* Bulletin officiel du Ministère de
                    l'Education Nationale n°31, 1992.
                    *Programmes de physiques et de chemie en classes de quatrième
                    et quatrième technologique.* Bulletin officiel du Ministère de
                    l'Education Nationale n°31, 1992.

Textbook   *Biology and Geology.* Nathan, 1992
                    Physical Science. Hachette, 1993
                    Sciences and Technology, C.M. Bordas, 1986

Germany

Curriculum Guide   *Richtlinien Biologie/Lernbereich Naturwissenshaften/Haupts-
                    chule.* Der Kulturminister des Landes Nordrhein - Westfalen,
                    1989
                    Richtlinien/Physik/Hauptschule. Der Kulturminister des Landes
                    Nordrhein - Westfalen, 1989
                    Richtlinien und Lehrpläne/Physik/Gymnasium II. Der

Kulturminister des Landes Nordrhein - Westfalen, 1981
Richtlinien und Lehrpläne/Biologie/Gymnasium/Sekundarstufe 1. Der Kulturminister des Landes Nordrhein - Westfalen, 1993
Richtlinien und Lehrpläne/Biologie/Gymnasium/Sekundarstufe 1. Der Kulturminister des Landes Nordrhein - Westfalen, 1981

Textbook
Elemente Chemie Nordrhein - Westfalen 7. Ernest Klett Verlag, 1988
Mensch und Raum: Geographie 7. Cornelsen Vergal, 1989
Physics für Gymnasien A1. Cornelsen Verlag
Terra, Geographie 7/8. Ernest Klett Verlag, 1988

Greece

Curriculum Guide
Curricula for Primary School Courses. O.E.Ø.B., 1987
Curriculum Guide of Lower Secondary School- Science. National Printing Service, 1985
Curriculum Guide of Upper Secondary School Science. National Printing Service, 1987

Textbook
Chemistry, 2nd Gymnasium Class (8th Grade). Organization for the Publication of Educational Books (O.E.Ø.B), 1991
Geography. Organization for the Publication of Educational Books (O.E.Ø.B), 1992
Human Biology 2nd Gynmasium (8th Grade). Organization for the Publication of Educational Books (O.E.Ø.B.), 1991
Physics. Organization for the Publication of Educational Books (O.E.Ø.B.), 1992
Physics, 2nd Gymnasium Glass (8th Grade). Organization for the Publication of Educational Books (O.E.Ø.B.), 1991
We and the World. Organization for the Publication of Educational Books (O.E.Ø.B), 1992

Hong Kong

Curriculum Guide
Syllabus for Physics (Advanced Level). The Curriculum Development Council, The Education Department, Hong Kong, 1992.
Syllabus for Science (Forms I-III). The Curriculum Development Council, The Education Department, Hong Kong, 1992.
Syllabuses for Primary Schools: Health Education. Government Printer, 1980
Syllabuses for Primary Schools: Primary Science. Government Printer, 1990

234 MANY VISIONS, MANY AIMS

Textbook	*Advanced Physics for Hong Kong. Volumes 1 and 2.* John Murray, 1992
	Further Physics. Volumes 1 and 2. Longman Group (Far East), 1992
	Health Education. Modern Educational Research Society Ltd
	Integrated Science Today Book 2. Jing Kung Educational Press, 1990
	New Geography 2 (Second Edition). Manhattan Press Ltd., 1990
	Patterns in Geography 2. Patterns, People and Places - Contrasting Environments. Jing Kung Educational Press, 1990
	Primary Science. The Art Publisher, 1988

Hungary

Curriculum Guide	*Biológia 6 -8. osztaly.* Országos Pedagógiai Intézet, 1978
	Fizika I - IV (A gimnáziumi oktatás és nevelés terre). Országos Pedagógiai Intézet, 1978
	Fizika 6 - 8. osztály. Országos Pedagógiai Intézet, 1979
	Körmyezetismeret 1-5. osztaly. Országos Pedagógiai Intézet, 1978
Textbook	*Biológia 8.* Tankönyvkiadó, 1985
	Fizika 8. Tankönyvkiadó, 1989
	Fizika IV. Tankönyvkiadó, 1991
	Földrajz 8. Tankönyvkiadó, 1992
	Kémia 8. Tankönyvkiadó, 1992
	Környezetismeret az általános iskolák 4. osztálya siámára. Tankönyvkiadó, 1988

Iceland

Curriculum Guide	*Aðalnámskrá Grunnskóla.* Menntamálaráðuneytið, maí, 1989
	Namskrá, Handa Framhaldsskólum. Menntamálaráðuneytið skólamálaskrifstofa, júní, 1990
Textbook	*Eðlis - og efnafræði Sérkenni Efna.* Námsgagnastofnun, 1983
	Eðlisfræði - fyrir framhaldsskóla 1B. IÐNÚ, 1990
	Eðlisfræði - fyrir framhaldsskóla 2. IÐNÚ, 1991
	Líf á Norðurslóðum - Inúítar. Námsgagnastofnun, 1985
	Líf í heitu landi - Tansanía. Námsgagnastofnun, 1977
	Lifandi verur - Fjölbreytni, flokkun, þróun. Námsgagnastofnun, 1981
	Líffélög og vistkerfi. Námsgagnastofnun, 1982
	Lífið í kringum okkur. Námsgagnastofnun, 1981

Við sjávarsíðuna Island. Námsgagnastofnun, 1982

Viðfangsefni og aðferðir. Námsgagnastofnun, 1981

Iran

Textbook

Experimental Sciences Grade 4. Ministry of Education, 1992

Experimental Sciences Grade 8. Ministry of Education, 1992

Physics. Ministry of Education, 1992

Ireland

Curriculum Guide

The Junior Certificate Science Syllabus. Department of Education, The Stationery Office, 1989.

Primary School Curriculum, Teacher's Handbook, Part 2. Department of Education, 1971

Rules and Programme for Secondary Schools. Department of Education, The Stationery Office, 1989 - 1991

Textbook

Discover Our World - Environmental Studies, Book 3. Folens, 1990

Earth Science - additional unit to "Science for the Future". Folens, 1991

Electronics - additional unit to "Science for the Future". Folens, 1991

Energy Conversions - additional unit to "Science for the Future". Folens, 1990

Food Science - additional unit to "Science for the Future". Folens, 1990

Horticulture - additional unit to "Science for the Future". Folens, 1991

Materials Science - additional unit to "Science for the Future". Folens, 1990

Science for the Future. Folens, 1990

Science From the Beginning. Oliver and Boyd; 13th impression, 1988

Senior Physics. Folens, 1987

Israel

Curriculum Guide

Biology Curriculum Guide for Senior High Schools. The Ministry of Education and Culture, 1991

Physics and Chemistry for Junior High Schools. The Ministry of Education and Culture, 1989

Physics for Senior High School. The Ministry of Education and Culture, 1980

Science in Technological Society. The Ministry of Education and Culture, 1988

	Textbook	*From Elements to Compounds.* Weizman Institute of Science, 1978

Textbook   *From Elements to Compounds.* Weizman Institute of Science, 1978

Issues on Reproduction. T.L. and Maaloth, 1991

It Burns/ Burning Materials. Ramot Publishing House, Tel Aviv University

Topics In Electromagnetism. Weizman Institute of Science, 1989

Topics in Modern Physics. Weizman Institute of Science, 1989

Waves and Physical Optics. Weizman Institute of Science, 1988

With the Current. 1992

Italy

Textbook   *Introduction to Experimental Sciences.* Le Monnier, April, 1980 - April, 1991

Laboratorio - Classe IV. Giunti-Marzocco, 1990

Laboratory for the New School. Giunti-Marzocco, 1992

Linee Chiare 4. Editrice La Scuola, 1990

Linee Chiare 5. Editrice La Scuola, 1990

Japan

Curriculum Guide   *The Course of Study for Elementary School Science.* The Ministry of Education, 1989

The Course of Study for Lower Secondary School Science. The Ministry of Education, 1989

The Course of Study for Upper Secondary School Science. The Ministry of Education, 1989

Textbook   *Physics.* Tokyo Shoseki Company, Ltd., 1993

Interesting Science, Grade 4, Volumes 1 and 2. Dai-Nippon Tosho Company, Ltd., 1993

Middle School Natural Science, First Field, Volumes A and B. Dai-nippon Tosho Company, Ltd., 1993

Middle School Natural Science, Second Field, Volumes. A and B. Dai-nippon Tosho Company, Ltd., 1993

New Science, First Field, Volumes A and B. Tokyo Shoseki Company, Ltd., 1993

New Science, Grade 4, Volumes 1 and 2. Tokyo Shoseki Company, Ltd., 1993

New Science, Second Field, Volumes A and B. Tokyo Shoseki Company, Ltd., 1993

Korea

Curriculum Guide   *National Curriculum Guide for Elementary School.* Ministry of Education, 1987

National Curriculum Guide for High School. Ministry of Education, 1988

National Curriculum Guide for Junior High School. Ministry of Education, 1987

Textbook

Chemistry. Kum Seung Publishing Co., 1989

Laboratory and Observation Workbooks 4-1 and 4-2. Ministry of Education, 1992

Nature Studies 4-1 and 4-2. Ministry of Education, 1992

Science 2. Dong - A Publishing and Printing Co., 1990

Physics. Dong-A Publishing & Printing Co., 1992

Latvia

Curriculum Guide

Basic Education Standards in Biology. Ministry of Education, Culture and Science

Basic Education Standards in Chemistry. Ministry of Education, Culture and Science

Basic Education Standards in Geography. Ministry of Education, Culture and Science

Fizikas Profilkuasa Vadlinijas. Ministry of Education, Culture and Science, 1995

Physics Standards. Ministry of Education, Culture and Science, 1992

Sakumizglitibas vadlinijal un standarti. Izglitibas Ministrija, 1992

Latvia

Textbook

Dabas maciba 4. Klasei. Zvaigue, 1989.

Fizikas Pamati. Zvaigue, 1992

Lithuania

Textbook

Biology: Animals (Grades 7-8). Prosveschenie, 1991

Chemistry: Inorganic Chemistry (Grade 8). Prosveschenie, 1989

Nature Study, Grade 3. Prosveschenie, 1991

Physics 8. Prosveschenie, 1991

Physics 11. Prosveschenie, 1991

Mexico

Curriculum Guide

ABC de Quimica. Segundo Curso Secundaria. Ediciones Numancia, S.A., 1991

Educación Básica, Primaria, Plan y programas de estudio de Ciencias Naturales y Geografía. Secretaría de Educación Pública, 1993

Educación Básica, Secundaria, Plan y programas de estudio de Ciencias. Secretaría de Educación Pública, 1993.

Programas maestros del tronco común del Bachillerato Technológico, Física. Consejo del Sistema Nacional de Educación Tecnológica, 1988

Textbook	*Ciencias Naturales Cuarto Grado.* Secretaría de Educación Pública, 1994
	Física 2. Publicaciones Culturales, 1994
	Física 3. Publicaciones Culturales, 1994
	Geografía Cuarto Grado. Secretaría de Educación Pública, 1994
	Geografía de México. Segundo Grado. Trillas, 1994
	La Magia de la Física. Ediciones Pedagógicas, 1994
	La Magia de la Química. Ediciones Pedagógicas, 1994
	Maravillas de la Biología 2. Ediciones Pedagógicas, 1994

Netherlands

Curriculum Guide	*Kerndoelen Basisonderwijs voor de natuur waaronder biologie.* Info-Reeks Basisivorming, PMB, 1993
	Kerndoelen Basisonderwijs voor natuur / scheikunde, biologie en aardrijkskunde, Inrichtingsbesluit vwo - havo - mavo - vbo. Info-Reeks Basisivorming, PMB, 1993
Textbook	*Biologie voor jou, 2mhv.* Malmberg, 1989
	Chemie, 3vwo / havo. Wolters - Noordhoff, 1989
	Geo Geordend 2hv. Meulenhoff Educatief, 1987
	Geoscoop, deel 2. Meulenhoff Educatief, 1990
	Natuurkunde in Contexten. Wolters - Noordhoff, 1989
	Natuurkunde voor nu en straks, 1hv. Thieme, 1983
	Natuurkunde voor nu en straks, 2lto. Thieme, 1979
	Natuurlijk!, deel 2. Malmberg, 1989
	Scheikunde mavo project, 3mavo Wolters-Noordhoff, 1981
	Scoop - Natuurkunde 5/6vwo. Wolters - Noordhoff, 1991

New Zealand

Curriculum Guide	*Science in the National Curriculum (Draft).* Ministry of Education, 1992
	Science Syllabus and Guide: Primary to Standard 4. Department of Education, 1989 (revised)
	University Bursaries/Entrance Scholarships, Physics in School Awards: Prescriptions. New Zealand Qualifications Authority, 1991
Textbook	*3 Science Books 1 and 2.* New House Publishers Ltd., 1989
	Senior Physics for Form Seven. Heinemann Education, 1985
Textbook Resource Booklets for Teachers	*Air.* New Zealand Dept. of Education, 1982
	Cells and Circuits. New Zealand Dept. of Education, 1980
	Discovering Electricity. New Zealand Dept. of Education, 1980
	Fungi. New Zealand Dept. of Education, 1982

Magnets. New Zealand Dept. of Education, 1980
Native Bush. Department of Education, 1980
Sandy Shore. Department of Education, Wellington, 1980
Solar System. New Zealand Dept. of Education, 1980
Sound. New Zealand Dept. of Education, 1980

Norway

Curriculum Guide   *Curriculum Guidelines for Compulsory Education in Norway.* The Ministry of Education and H. Aschehoug & Co., 1987
Fysikk. Ministry of Education and Gyldendal, 1992

Textbook   *Arbeidsbok til ìVerden For Alleî.* Universitetsforlaget, 1988
Forsøk og Fakta. NKS - forlaget, 1989
Heimkunnskap. Det Norske Samlaget, 1989
Rom, Stoff, Tid. 3 FY Cappelen, 1991
Rom, Stoff, Tid. 3 FY (laboratory manual), Cappelen, 1991
Verden For Alle, Faktabok. Universitetsforlaget, 1987
Verden For Alle, Undrebok. Universitetsforlaget, 1987

Philippines

Curriculum Guide   *Desired Learning Competencies.* Bureau of Secondary Education, Department of Education Culture and Sports, 1991

Textbook   *Science and Health for a Better Life.* Department of Education Culture and Sports, 1985
Science and Technology I. Department of Education Culture and Sports, 1989

Portugal

Curriculum Guide   *Ensino Basico - Programa do 1º Ciclo.* Direcçáo Geral do Ensino Básico e Secundário. Ministério da Educação, Editorial do M.E. - Algueirão, 1990
Ensino Basico - Programa do 1º Ciclo. Direcçáo Geral do Ensino Básico e Secundário. Programa de Física e Química , Editorial do M.E. - Algueirão, 1992
Organização Curricular e Programas (Volume 1) Ensino Básico, 3.º Ciclo. Direcçáo Geral do Ensino Básico e Secundário. Ministério da Educação, Editorial do M.E. - Algueirão, 1990

Textbook   *Física 8.º.* Didáctica Editora, Lisboa, 1994
O Bambi 4, Estudo do Meio. Porto Editora, Porto, 1994
Química 8.º. Didáctica Editora, Lisboa, 1994
Vida Humana - Ciências Naturais 8.º Ano. Porto Editora, Porto, 1994

Romania

Curriculum Guide

Programa de biologie. Editura Didactică si Pedagogică, 1991

Programa de biologie pentru invatamantul primar. Editura Didactică si Pedagogică Bucuresti, 1991

Programa de chimie- pentru invatamantul gimnazial. Edutura Didactică si Pedagogică , 1991

Programa de fizica pentru gimnaziu. Editura Didactică si Pedagogică Bucuresti, 1991

Programa de fizica pentru liceu. Editura Didactică si Pedagogică Bucuresti, 1991

Programa de geografie pentru clasele V-VIII/5-8. Editura Didactică si Pedagogică Bucuresti, 1993

Textbook

Anatomia Fisiologia si Igieno Omului. Editura Didactică si Pedagogică Bucuresti, 1991

Biologia. Editura Didactică si Pedagogică Bucuresti, 1992

Chimia. manual pentru clasa a VIII- a. Editura Didactică si Pedagogică Bucuresti, 1992

Chimie. Manual pentru clasa a XII-a, Editura Didactică si Pedagogică Bucuresti, 1992

Cunoasterea Mediului Înconjurător. Manual pentru clasa a IV-a. Editura Didactică si Pedagogică, R.A., 1992

Fizica. Manual pentru clasa a VIII- a. Editura Didactică si Pedagogică Bucuresti, 1992

Fizica. Manual pentru clasa a XII-a. Editura Didactică si Pedagogică Bucuresti, 1992

Geografia Romaniei. Manual pentru clasa a VIII-a/8. Editura Didactică si Pedagogică Bucuresti, 1992

Geografie. Manual pentru clasa a XII-a. Editura Didactică si Pedagogică Bucuresti, 1992

Russian Federation

Curriculum Guide

Curricula of General Secondary School Biology. Ministry of Education, 1990

Curricula of General Secondary School Chemistry. Ministry of Education, 1990

Curricula of Physics. Ministry of Education, 1990

Curricula of Primary School (Grades 1-3) Nature Study. Ministry of Education, 1990

Curricula of Secondary General School. Physics. Astronomy. Ministry of Education, 1990

Textbook   *Biology: Animals (Grades 7-8).* Prosveschenie, 1991
                            Chemistry: Inorgani chemistry (Grade 8). Prosveschenie, 1991
                            Nature Study, Grade 3. Prosveschenie, 1991
                            Physics 8. Prosveschenie, 1991
                            Physics 11. Prosveschenie, 1991

Scotland

Curriculum Guide   *Environmental Studies 5 - 14 (Draft).* Scottish Office Education
                            Dept., 1992

Textbook   *Learning through Science: Colour.* Macdonald Educational/
                            Simon and Schuster, 1982
                            Learning through Science: Electricity. Macdonald Educational/
                            Simon and Schuster, 1982
                            Learning through Science: Materials. Macdonald Educational/
                            Simon and Schuster, 1982
                            Learning through Science: Moving Around. Macdonald
                            Educational/Simon and Schuster, 1982
                            Learning through Science: Out of Doors. Macdonald
                            Educational/Simon and Schuster, 1982
                            Learning through Science: Time, Growth and Change.
                            Macdonald Educational/Simon and Schuster, 1982
                            Starting Science. Oxford U Press, 1986

Singapore

Curriculum Guide   *GCE Advanced Level Syllabus (Physics).* University of
                            Cambridge, Locals Exams Syndicate International Exams
                            Science Syllabus (Lower Secondary). Curriculum Planning
                            Division, Ministry of Education, 1990
                            Science Syllabus (Primary). Curriculum Planning Division,
                            Ministry of Education, 1990

Textbook   *Advanced Level Physics (6th Edition).* Heinemann, 1987
                            Lower Secondary Science Project Exploring Science 2.
                            Curriculum Development Institute of Singapore (CDIS), 1984
                            Primary Science Book 4. Curriculum Development Institute of
                            Singapore (CDIS), 1982

Slovak Republic

Curriculum Guide   *Ucebné osnovy 2. stupna základnej skoly. Chémia (7.-8. rocník).*
                            Slovenské Pedagogické nakladatelstvo, 1987
                            Ucebné osnovy 2. stupna základnej skoly. Fyzika (6.-8. rocník).
                            Slovenské Pedagogické nakladatelstvo, 1987
                            *Ucebné osnovy 2. stupna základnej skoly. Prírodopis (5.-8. roc-
                            ník).* Slovenské Pedagogické nakladatelstvo, 1987

Ucebné osnovy pre gymnasia. Fyzika (1.-4. rocník). Schválilo Ministerstvo skolsta, mládeze a telesnej vychovy Slovenskej republiky dna, 1990

Textbook   *Chémia pre 8.rocník.* Slovenske Pedagogicke Nakzadatelstvo, 1992

Fyzika pre 8. rocník základnej skoly A. Slovenske Pedagogicke Nakzadatelstvo, 1992

Fyzika pre 8. rocník základnej skoly B. Slovenske Pedagogicke Nakzadatelstvo, 1992

Prírodopis 4. Fyzika pre 4 rocník Gymnazit. Slovenske Pedagogicke Nakzadatelstvo, 1991

Prírodopis 8. Slovenske Pedagogicke Nakzadatelstvo, 1991

Slovenia

Curriculum Guide   *Obvezni predmetnik in učni načrt osnovne sole.* Zavod Republikes za Slovenije Solstvo in šport, 1983

Textbook   *Fizika.* Državna založba Slovenije, 1991

Fizika 8. Državna založba Slovenije, 1992

Organska kemija. Državna založba Slovenije, 1993

Razvojni nauk. Državna založba Slovenije, 1988

Spoznavanje narave za 4. razred. Državna založba Slovenije, 1993

South Africa

Curriculum Guide   *Core Syllabus For Geography Ordinary Grade Standards 5,6,7.* Department of Education and Culture

Final Core Syllabus for Physical Science (Higher Grade) Secondary School Standards 8, 9 and 10. Department of Education and Culture,1982

Syllabus for Environment Studies. Department of Education and Culture

Syllabus for General Science, Standards 2, 3 & 4. Department of Education and Culture

Syllabus for General Science, Standards 5, 6, 7. Department of Education and Culture

Textbook   *Active General Science Standard 6.* DeJager-HAUM Publishers, 1984

Employment: Jobs done by People. (Pre-publication edition).

Enjoy Geography Standard 6. Lexicon Publishers and Hodder & Stoughton, 1992

Exciting Geography. Lexicon, 1985
General Science in Action 6. Juta and Co, Ltd., 1984
Our Environment. Environment Studies for Standard 2. (Manuscript for translations into African languages.)
Physical Science in Action: Standard 10 (HG/SG). DeJager-HAUM, 1986
Successful Science 10. Oxford University Press Southern Africa, 1989
The World Around us 2. Juta and Co, Ltd., 1980

Spain

Curriculum Guide

Area de Ciencias de la Naturaleza. Vida Escolar, 1978
Programación Física COV. B.O.E., 17-Marzo, 1978
Programas renovados del Ciclo Medio. Vida Escolar, 1982

Textbook

Espora, Ciencias Naturales. Anaya, 1990
Espora, Ciencias Naturales. 8º. Anaya, 1986
Física Energía C.O.U. Ediciones SM, 1992
Naturaleza 4 EGB Ciclo Medio. Santillana, 1992
Nova 8, Naturaleza EGB. Santillana, 1992

Sweden

Curriculum Guide

Läroplän för Grundskolan (Lgr 80). Liber Utbildningsförlag, Skolöverstyrelsen, 1990
Läroplän för Gymnasieskolan (Lgy 70) Supplement 32. Skolöverstyrelsen och Utbildningsförlaget 1981

Textbook

Biologi för grundskolans högstadium. Gleerups, 1994
Fysik 90 Del 1. Almgvist & Wiksells Förlag AB, 1989
Fysik för Gymnasieskolan. Årskurs 3, Natur och Kultur, 1993

Switzerland

Curriculum Guide

Curriculum Guides for the Lower Secondary School. Dipartimento della pubblica educatione, 1987
Curriculum Guides for the Primary School. Dipartimento della pubblica educatione, 1984
Curriculum Guides for the Upper Secondary School. Dipartimento istrutione e cultura, 1992
Lehrplän für die Primarschulen des Kantons Luzern. Ausgabe, 1984
Lehrplän für die Volksschule des Kanton Zürich. Erziehungs-direktion des Kanton Zürich, 1991
Lehrplän Geographie, Primarschule/Sekundarschule, 5.-9. Schuljähr. (Bern Government), 1983

Lehrplän Heimatunterricht, Primarschule 1.-4. Schuljähr. (Bern Government), 1983

Lehrplän Natürkunde, Sekundarschule, 5.-9. Schuljähr. (Bern Government), 1983

Naturlehre. Kommission Naturlehre IEOK, 1991

Physik. Programm mit Kommentar. VSMP, 1976

Plan d'études romand pour les classes de 1re à 6e année. Secrétariat à la coordination romande, 1989

Plan d'études: Sciences expérimentales, travaux pratiques de biologie. Ministère de l'éducation (Juras), 1993

Textbook   *Bau und Funktionen unseres Körpers.* Lehrmittelverlag des Kantons Zürich, 1989

Geographie Europas. Lehrmittelverlag des Kantons Zürich, 1991

Invito alla fisica 2. Zanichelli Editore S.P.A., 1991

Invito alla fisica 3. Zanichelli Editore S.P.A., 1991

Knowing the Environment. Third Primary. Department of Public Instruction of Vallis, 1985

Pflanzenkunde. Verlag Paul Haupt Bern, 1985

Physics. Experimental Science. Chemistry. L.E.P., 1992

Physik. Schroedel Schulbuchverlag, 1989

Tierkunde. Verlag Paul Haupt Bern, 1981

Welt der Physik und Chemie. Schroedel Schulbuchverlag, 1991

Tunisia

Textbook   *Sciences Naturelles. Deuxième année de l'enseignement secondaire.* Centre National Pédagogique, 1992

United States

Curriculum Guide   *Chapter 75 Curriculum- for Grades K-12.* Texas Education Agency, 1992

The Chemistry of Matter - Block H. The University of the State of New York. The State Department of Education, 1988

Core Course Proficiencies: Science. New Jersey Department of Education, 1990

Curriculum Frameworks for Grades 6-8 Basic Programs-Science and Health (Florida). Florida Department of Education

Elementary Science Course of Study and Curriculum Guide. Idaho State Department of Education, 1989

Elementary Science Curriculum Guide. New Jersey Department of Education, 1985

Elementary Science Syllabus. The University of the State of New York. The State Department of Education, 1992

Energy: Sources and Issues - Block I. The University of the State of New York. The State Department of Education, 1991

Energy and Motion - Block G. The University of the State of New York. The State Department of Education, 1990

Essential Content. Hawaii Department of Education, 1992

A Guide to Curriculum Planning in Science. Wisconsin Department of Public Instruction, 1990

Guidelines for Science Curriculum in Washington Schools. Office of the Superintendent for Public Instruction, 1992

Learner Outcomes - Science. Oklahoma State Department of Education

Quality Core Curriculum Guides for K-12 Students in Georgia.

Science. State of Nevada, Department of Education

Science Comprehensive Curriculum Goals- a Model for Local Curriculum Development. Oregon State Department of Education, 1989

Science Curriculum Guide- Section 3, Science. Idaho State Department of Education, 1989

Science Framework for Grades K-12. California State Board of Education, 1990

State Goals and Sample Learning Objectives- Biological and Physical Sciences for grades K-12. Illinois State Department of Education

Weather and Climate - Block E. The University of the State of New York. The State Department of Education, 1988

Wyoming Standards of Excellence for Science Education. Wyoming Department of Education, 1989

Textbook   *Earth Science.* Merrill, 1993

HBJ Science- for grade 4. Harcourt Brace Jovanovich, 1989

Life Science. Merrill, 1993

Modern Physics. Holt, Rinehart and Winston, 1992

Physical Science- for grade 8. Merrill, 1993

Science Horizons. Silver Burdett and Ginn, 1991

Science in Your World. McMillan / McGraw Hill, 1991

University Physics. Addison Wesley, 1992

Appendix H

ACKNOWLEDGMENTS

A project of this scope, magnitude, and complexity only comes about through the concerted efforts of many people. And so many have made valuable contributions at various times over the 4-year span of this project that any list runs the risk of omitting someone. Nonetheless, the important roles a number of people have had in the creation of this report need to be acknowledged. While the substance of this volume reflects the perspective of the authors alone, this publication would not have been possible without the participation of the following people:

- Al Beaton, the Third International Mathematics and Science Study (TIMSS) director, has provided general oversight and direction to the many different components of the TIMSS research project.

- Tjerd Plomp, chair of the TIMSS Steering Committee and president of the International Association for the Evaluation of Educational Achievement (IEA), provided guidance and IEA financial support so some countries could send representatives to the curriculum analysis training.

- Leigh Burstein, Curtis McKnight, Senta Raizen, David Robitaille, William Schmidt, David Wiley, and Richard Wolfe developed the conceptualization of the curriculum analysis.

- Curtis McKnight, Edward Britton, Senta Raizen, David Robitaille, William Schmidt, Jane Swafford, and Geoffrey Howson wrote the first drafts of the curriculum frameworks.

- Curtis McKnight, Edward Britton, William Schmidt, and Gilbert Valverde developed the curriculum analysis training and quality control procedures and conducted all the training sessions for those from the participating countries.

- Kjell Gisselberg (Stockholm), Derek Gray (Pretoria), Einar Gudmundsson (Reykjavik), Milagros Ibe and Ester Ogena (Quezon City), Doris Jorde and Svien Lie (Oslo), Frederick Leung (Hong Kong), William Loxely (the Hague), Eduardo Luna (Miami), Jahja Umar (Djokjakarta), and Pavla Ziekniecova (Prague) organized and hosted curriculum analysis training sessions.

- Gilbert Valverde coordinated quality control, communicated with countries, and led the quality control team of Peter Achuonjei, Jimin Cho, Suwanna Eamsukkawat, Virginia Keen, Zora Mohammadi-Ziazi, Yu Qi, Charles Rop, and Wen-Ling Yang as they spent long hours entering data and enacting the extensive quality control procedures.

- Leonard Bianchi, Richard Houang, Curtis McKnight, William Schmidt, Gilbert Valverde, David Wiley, and Richard Wolfe designed the analyses.

- Leonard Bianchi and Richard Houang designed and constructed the curriculum analysis database.

- Leonard Bianchi led the statistical production team of Haiming Hou and Wen-Ling Yang, Christine deMars, and Shelly Naud in writing analysis programs and producing the tables and displays that supported the authors' efforts.

- Nancy Law spent 3 months at the Curriculum Analysis Center carefully examining the Hong Kong data and provided valuable contributions to the international data analyses.

- Gordon Ambach, Emilie Barrier, Paul Black, Rolf Blank, Antoine Bodin, Vladimir Burhan, John Dossey, Robert Garden, Ignacio Gonzalo, Jeanne Griffith, Geoffrey Howson, Doris Jorde, Jeremy Kilpatrick, Christine Keitel-Kreidt, Peter Labudde, Nancy Law, Marcia Linn, Robert Linn, Mick Martin, Urs Moser, Graham Orpwood, Eugene Owen, Lois Peak, Richard Prawat, Miriam Reiner, David Robitaille, Paul Sally, Jack Schwille, Ramsey Seldon, Katsuhiko Shimizu, Elizabeth Stage, Larry Suter, James Taylor, Kenneth Travers, Richard White, and Theo Wubbels all provided advice at key times in the project's evolution.

- The TIMSS' Advisory Committee on Curriculum Analysis (ACCA) – Fernando Córdova Calderón, Galina S. Kovalyova, Wilmad Kuiper, Frederick Leung, and Anne Servant – provided comments, helped in the design of the analyses underlying this volume, and reviewed initial drafts.

- Jacqueline Babcock, Nikki Butler, Leland Cogan, Emilie Curtis, Jason Durr, Steve McConnell, Mike Reed, Sudip Suvedi, and Lorene Tomlin provided expert and tireless support in the many mundane and tedious technical aspects of producing a publication.

Although too numerous to mention by name, the enormous effort and careful documentation of those in each country who mastered the TIMSS curriculum frameworks and coded all the curriculum guides and textbooks must be gratefully acknowledged. These people devoted a large amount of energy, time, and dedication to implement our vision of a substantive, international analysis of curricula. This volume would not have been possible without them. We hope they will be able to take pride in this first fruit of our labor.

Appendix I

TABLES AND FIGURES FOR ALL PARTICIPATING COUNTRIES

Figure 13.2

Textbook Space Devoted to Major Science Topics.

This figure shows the percentages of textbook blocks devoted to the eight broadest science framework content categories for the textbooks of the upper grade of Population 2 in all countries.

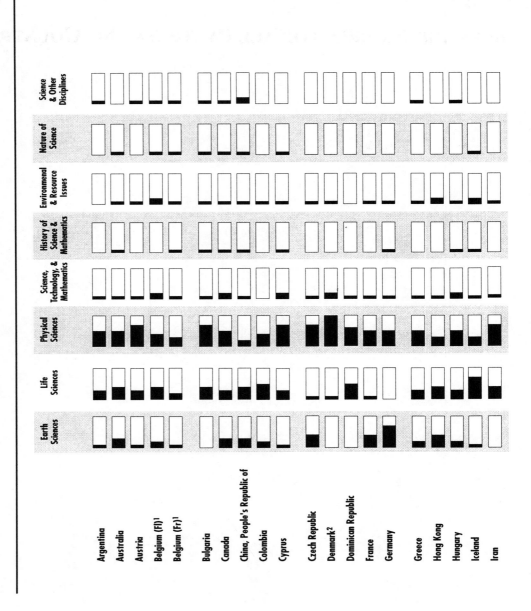

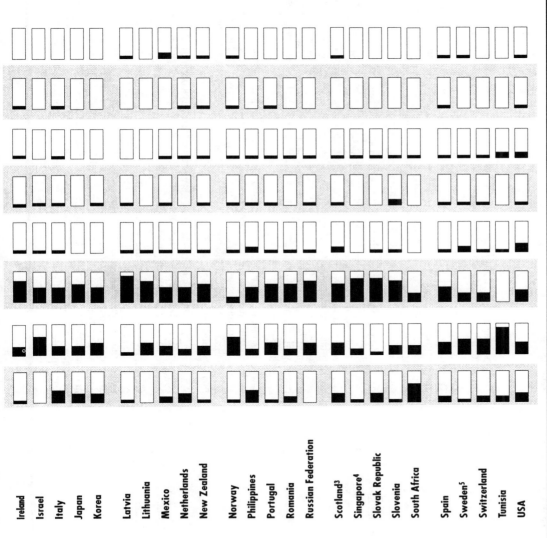

Legend:

0%	21-30%	51-60%	81-90%
1-10%	31-40%	61-70%	91-100%
11-20%	41-50%	71-80%	

Ireland
Israel
Italy
Japan
Korea

Latvia
Lithuania
Mexico
Netherlands
New Zealand

Norway
Philippines
Portugal
Romania
Russian Federation

Scotland[3]
Singapore[4]
Slovak Republic
Slovenia
South Africa

Spain
Sweden[5]
Switzerland
Tunisia
USA

1   The National Research Coordinators of Belgium have collected data only from curriculum guides. Due to the great level of detail of the guides and
    their extensive use, data from these are compared in this display with the textbook data supplied from all other countries

2   Denmark only provided data collected from physical science textbooks for this population.

3   In Scotland, textbooks receive only limited use in science education. Teachers are likely to use a variety of resources. The Scottish Curriculum for
    science education includes a variety of topics outside the TIMSS science frameworks subsumed under the title "Environmental Studies."

4   The Curriculum Guide for Singapore was revised in 1992. The National Research Coordinator estimates that in the revised textbook the percent-
    ages devoted to "Physical Sciences" and "Life Sciences" have changed to 64% and 30% respectively.

5   Although chemistry books do exist at this level, none were coded.

Table 13.3

Summary of Science Reform Interests in TIMSS Countries.

Widespread interest was expressed in science education reforms. Here is a summary of reform interests in TIMSS countries as reported by experts within each country.

Country	Any initiatives in last 5 years?	Remarks
Argentina	—	Textbooks are being used less due to educational policies prohibiting teachers to demand any textbooks for students. The textbooks stress applications and have explanatory notes.
Australia	Yes	There is a greater emphasis on the relevance of science and the applications of science in everyday life. The human aspects of science and the impact of science on the environment and society are an integral part of new syllabuses. There is a corresponding change in the methodology of science teaching and learning related to practice and introduction of concepts linked to students' experiences and real world contexts. Programmable and graphic calculators, computers, and interface devices are used for measurement, analysis, and presentation of data.
Belgium (Flemish-speaking & French-speaking)	Yes	An important innovation is that much has been accomplished in terms of 'quality control' in Belgian education. Final objectives have been developed for science. These set the minimum standar for what has to be learned. Selected issues include: (1) more attention will be given to the learning of concepts. Graphical solutions will be stressed and more audiovisual means will be introduced; (2) a less encyclopedic approach to the teaching of physics will be stressed; (3) in biology, a more experimental and practical approach will predominate and an introduction to an ecological view of the living world will be provided; (4) minimum equipment facilities for science classes will be stipulated; and (5) although the use of computer applications in the diverse fields of science is still limited, attention will be given to technological developments in this area.
Bulgaria	Yes	There is some experimentation to integrate science in middle schools, introduce environmental studies, and differentiate teaching according to student interests and abilities. New trends include integrated science in secondary schools and the reduction of material covered. The use of calculators and computers in schools has increased in the past 10 years and software has been introduced for the teaching of chemistry, physics, and biology.
China, People's Republic of	Yes	Additions to the curriculum include energy resources, environment, population issues, pollution, electronic technology, computers, and space technology. New technologies in use include microcomputers, VCRs, TVs, projecting apparatus, etc.
Canada	Yes	Generally there has been an increased emphasis on: (1) understanding scientific processes and principles, and problem solving; (2) conceptual understandings approached from a phenomenological basis rather than from a mathematical basis, particularly in physics; (3) integration within the sciences -- there are usually separate courses for biology, chemistry, etc., only in senior secondary; (4) integration of sciences with other subject areas often centered on "themes"; (5) issues surrounding the impact of science and technology on society; (6) teaching from constructivist principles; (7) awareness and promotion of science-related careers; (8) issues of gender equity and cultural bias; and (9) development/evaluation of appropriate scientific software. There is a general acceptance of calculator and computer use by students at all levels. New Science/Technology/Society courses have been introduced.

Country		
Cyprus	Yes	Attempts are being made to identify broader topics such as environment, health education, and ecological issues. Emphasis is on method rather than content with a more hands-on approach. Efforts are underway to provide schools with more science equipment. There is more emphasis on societal issues of science. The concept of project work has been introduced in the science curriculum. Technology has been introduced in secondary schools. There has been no essential change in curriculum although a group of science teachers is working toward the development of materials for teachers of science. The Human Biology textbook is based on a series of videodisks, which is available to some teachers.
Czech Republic	Yes	Curriculum content has been gradually reduced while preserving the concepts that enable a broader region of facts to be described. There is increased attention paid to the effects of technical processes on the environment, both locally and globally, but this has had little effect on curricula. The daily use of calculators, computers, and video-equipment is recommended.
Denmark	Yes	The following is based on 1988 guidelines. *Physics:* Students must be able to interpret phenomena in thier immediate surroundings from a physics point of view; must acquire the idea of physics as a coherent description of nature; must learn the history and theory of physics and its applications; must acquire an insight into the close relationships between progress in physics and the development of society and technology. Use of calculators in physics instruction has been permitted since 1976 but is not allowed at written examinations. *Chemistry:* Students must learn the methods and applications of chemistry in everyday life; must be acquainted with publications from environmental organizations. Computers are used for collecting and processing chemical test results as well as for searching external databases, writing, and calculating. *Biology:* There has been a shift of focus to analytical problem-oriented teaching as well as new theoretical and practical knowledge and research. New topics include disease prevention, biological engineering, medical technology, production technology, etc. A new act affecting the Folkeskole came into force in 1994. The new curriculum guides set new goals and objectives for all science subjects, and are structurally different from the ones used in the present TIMSS study. Additionally, a new science subject has been introduced in grades 1 through 6: "Nature and Technology." Furthermore, it is stressed that "a green weft" is to be in all subjects in the Folkeskole, not only science.
France	Yes	In recent years, especially in biology and geology, more importance is given to the teaching through problems around concepts. The general approach is more naturalistic and less conceptual. There has been an increased emphasis on technical aspects; on science's everyday applications; and a more systematic approach to the study of life processes (e.g., human nutrition). In physics, changes are motivated by the desire to prepare pupils to interact efficiently with a technological world. Since 1992, physics and chemistry begin at grade 8 instead of grade 6. There is a concern to develop a critical mind, integrate new technologies and computers, and, consequently, individualize learning. There has been an introduction of computer science and computer-assisted experimentation in both biology and physics. Software is making the learning of some theoretical concepts easier, allowing the teaching of physics in some additional fields. New programs have been or will soon be written for grades 8 through 12.
Germany	No	Much is going on in the area of research and development, which is affecting teacher training. The computer is becoming increasingly important. The use of new technologies has been included in the core curriculum either as a separate unit or incorporated in the mathematics syllabus.
Greece	—	The major change in science over the past 10 years is the shift from learning facts to understanding concepts and their application to everyday life. Observations and experiments are encouraged but are not done widely due to lack of resources and poor teacher training. There is no link between science curriculum and use of new technologies.

Table 13.3 (continued)
Summary of Science Reform Interests in TIMSS Countries.

Country	Any initiatives in last 5 years?	Remarks
Hong Kong	Yes	Primary level science was introduced in 1984. High school science courses are expected to be revised in the next 5 years. There are no changes proposed for primary or junior secondary science. Technology is recognized mainly as an extension of woodwork and metalwork and is now called Design and Technology at grades 7 through 9.
Iran	No	About 3 years ago, a science curriculum committee formulated new science education goals and curriculum frameworks for both primary and secondary education levels. The committee's main points are that new topics of science and technology need to be incorporated into textbooks and that most teaching methods are old and inefficient. Use of computer technology is rapidly increasing in all educational settings.
Ireland	Yes	*Primary School.* The emphasis on science is minimal even in the "intended" curriculum. Nature Study and Elementary Science compete for only 15 percent of the weekly schedule. The open nature of teacher guidelines does allow individual teachers to treat the subject more or less as they wish. *Junior Certificate Level.* The new science syllabus emphasizes science as a practical subject with everyday applications with enormous impact on society. New units have been introduced in the areas of applied science and local studies. Technology has been introduced as a new subject with focus on crafts and materials and options for energy, the media, communications and industry. *Leaving Certificate Level.* New courses in chemistry and physics were introduced in 1983 and 1984 with more emphasis on the practical nature of these subjects and their relevance to industry. Laboratory work became mandatory. Calculators are not admissible for examinations so are not encouraged in the classroom. *Senior Level.* Science is split into the three traditional branches of physics, chemistry, and biology. Environmental science was introduced in the chemistry course in 1983. Calculators are used at this level. The use of computers as a laboratory resource is considered a specific objective in physics.
Israel	Yes	Elementary science has been changing from pure science to emphasis on application and the discovery approach to the learning of concepts. In high school biology, the teacher chooses to teach six out of nine possible topics. A new curriculum on environmental studies is available, but few schools offer it. "Introduction to the Microcomputer" is offered in most junior high schools. Many high schools offer courses in computer literacy.
Japan	Yes	National courses of study were revised in 1989 and are not expected to be revised for another 10 years. The biggest change in the last 10 years is a shortening of the number of hours for science. The relationship between study and daily life is gaining greater importance at all levels. In elementary schools, science is included in "life environmental studies," which was just added to the curriculum. In lower secondary, elective science courses were added and the use of computers was introduced. Students study about the computer as a tool of information science, the computer's development, and development of computer chips. In upper secondary, there has been a tendency to make contents of science easier.

Country		Description
Korea	Yes	A new unit on "Environment and Natural Resources" was introduced in the high school earth science curriculum. Content includes "human and natural environment, natural resources of the Earth, and the future Earth." Proposed is a new integrated science curriculum called "General Science" for tenth graders with a greater emphasis on environmental concerns and technology. STS education and inquiry activities within everyday life contexts are emphasized at the primary and secondary school levels. The number of computers and VCRs in schools in growing, but few classes use them for instruction.
Latvia	Yes	The curriculum has been continually changing with an emphasis on real-world and environmental problems.
Lithuania	Yes	The reform in science education began in 1989 when some new integrated science and civic education courses were developed for primary school (grades 1 through 4). The plan is to extend integrated science courses to grades 5 through 7 leaving specific science subjects (i.e. physics, chemistry, biology, geography) for grades 8 through 12. The number of topics has been reduced with the focus being scientific literacy, especially that related to environment, human biology and health, and new technologies. Methods of active teaching are being stressed. Some new textbooks and teaching materials were prepared by Lithuanian authors; others were translated from the Western countries. The reform is linked with rather intensive teacher training activities at the secondary level where the goals and objectives of reform in science education are discussed and some new methods of teaching are demonstrated.
Mexico	No	The current science curriculum was implemented in 1974 with changes expected in 1992 -- there will be a greater focus on applications and the problem solving process. New technologies have not been introduced in all secondary schools. The teaching of science has not changed due to technological applications.
Netherlands	Yes	Characteristics of changes in primary science education are: (1) a broad and integrated conception of elementary science content including biology, physics, chemistry, earth science, environmental and technological education; (2) the importance given to the immediate surroundings of the school as a source for science teaching and learning; and (3) an emphasis on the importance of hands-on activities. The Dutch system has four tracks in secondary education: pre-university (VWO, a 6-year program), senior general secondary education (HAVO, a 5-year program), junior general secondary education (MAVO, 4 years), and junior secondary vocational education (VBO, 4 years). In 1993, a core curriculum of 15 subjects was implemented in the first 3 years of all four tracks. Basic characteristics of these new intended curricula are: (1) the selection of science content appealing to students, the use of real-life contexts form learning science so that students apply science in a creative and meaningful way; (2) an emphasis on activity-based student engagement, aimed at the acquisition of a rich variety of investigative, social, and communication skills like conducting experiments, organizing and representing data, applying concepts, using the computer; (3) coherence across science topics and between science and other subjects including mathematics. In upper secondary education, national exams for elective courses in physics, chemistry, and biology (VWO, HAVO) have been implemented over the past 10 years. These programs are still constructed around the traditional structure. A substantial overall reform of the last two grades of HAVO and the last three grades of VWO is currently being prepared and their implementation are envisaged for 1998.

Table 13.3 (continued)
Summary of Science Reform Interests in TIMSS Countries.

Country	Any initiatives in last 5 years?	Remarks
New Zealand	Yes	In the past decade, curriculum development has been limited to middle and senior schools. Several senior secondary programs were reviewed: biology; chemistry (1986), which resulted in a change to the chemistry examination prescriptions and consequently a more theoretical approach to teaching; and physics (1990), which gave rise to a more practical, process-orientated course. A review of science in the final years of primary school and junior secondary school resulted in the publication, in 1989, of a discussion document. Although it never became an official document, it did have an effect on the implemented curriculum. In 1992, Ministry of Education officials and teaching professionals completed the development of a new national science curriculum for all levels of the school system. Its implementation began in 1995. This recent science development attempts to achieve "science for all": science is now offered as an integrated discipline at the senior secondary level, as well as discrete disciplines. Broad skills such as investigating, problem solving, and cooperative learning are emphasized, moving away from the traditional "following recipes" method of experimentation. The place of science and technology in society is emphasized. The development also attempts to address issues of low participation in science at senior levels by females and ethnic minority students. The use of computers, digital sensors, and other equipment is increasingly becoming part of the student investigation work.
Norway	Yes	Subject matter is now presented for 3-year periods. There is greater freedom for schools and individual teachers to draw up their list of specific topics to be taught. Sciences and social sciences are combined into one subject called "civics" grades 1 through 6. There is more emphasis on general concepts and scientific processes, the environment, and pollution. In upper secondary there were smaller changes – science courses are more focused on the environment. Computers have had minimal impact.
Portugal	Yes	The expansion of technical knowledge has been bringing on changes in all other areas of the curriculum. Many curriculum topics are aimed at developing a scientific attitude. Contents are organized around themes and conceptual frameworks. New content includes the impact of science and technology on society. Computers and calculators are not considered in the intended curriculum. Computers were introduced in Portuguese schools in 1986 and 1987 through a nationwide project called Project MINERVA. Problem solving activities are increasing in use.
Romania	No	An attempt has been made to eliminate certain subjects due to the large quantity of material covered by each discipline. Practical applications have been reduced.
Russian Federation	Yes	Changes during the past 10 years didn't involve curriculum structure. The change has been from mastering knowledge and skills to developing learning skills. Environmental issues have been given greater attention. The introduction of 4 years of primary schooling (instead of 3) in 1988 did cause structural changes in the elementary science curriculum - two new courses were added: "Surrounding World" for

Scotland	Yes	grades 1 to 2 and a "Nature Study" for grades 3 to 4. Due to social changes the entire system of public education is more democratic and humanistic. Two divergent science programs have been proposed for the middle school: the study of all science subjects separately in grades 5 through 9; an integrated science program in grades 5 through 7 with differentiation in grades 8 to 9. In upper secondary school a basic level of science courses is provided and advanced courses are offered for students who choose programs in physics-mathematics, science or technical studies. A new course "Informatics" has been added to the curriculum and schools are being supplied with computers. However, most schools don't have enough computers and calculators so there are no concrete recommendations for their use in science instruction. Materials describe general trends in technology in connection with societal problems, primarily because technical descriptions change so quickly.
Singapore	Yes	New national guidelines include energy and forces, earth and space, living things, and the processes of life. They include references to the impact of science and technology on society. SG sciences: four courses are divided into a number of topics that reflect major subject themes. HG sciences: recently reduced historical elements to increase relevance. A new course in human biology gives attention to environmental issues and behavior. CSYS courses: allow students to specialize by studying 2-3 topics in greater depth. Students carry out investigations and work with highly technological equipment. There are also modular courses to accommodate a greater number of students returning to senior school focusing on environmental issues, resource management, electronics, etc. Computers are available in nearly all science departments and are used for a variety of purposes. There is limited use of videodisks. Graphics calculators are often required in advanced mathematics courses and are used in physics courses at the higher grades. All courses include some mention of subject applications.
Slovak Republic	Yes	There is more emphasis on content knowledge acquisition as well as process skills and attitude development in the past 10 years. Schools are encouraged to incorporate practical tests and project work to assess skills and attitudes. Secondary School: the emphasis is on understanding and application of content knowledge and use of process skills such as interpretation of data and problem solving. New topics relevant to a technical society have been introduced such as electronics, human health, food technology, and the impact of man on the environment. As the accessibility to computers increases, some teachers are beginning to use them in instruction, remedial work, and enrichment.
		Primary School. There has been a reduction of biology content. Health and environmental studies have been introduced. Physics is based on four integrative concepts: particle structure of substances, physical field, energy, and power. Demonstrations and experiments by students have been incorporated into the curriculum and teachers are required to teach them. *Secondary School.* Education goals have been changed to emphasize not only knowledge of facts but also the development of intellectual and manual skills: laboratory work, creativity, independent thinking, interpretation of observed phenomena and experimental results. Biology includes several new topics: physiology, molecular biology, genetics, ecology, and etiology. The number of lessons allotted to biology has increased and about one-third of this time is spent in practical laboratory work. The traditional structure of physics content has been replaced by a system of integrated knowledge (forms and reasons of motion, structure and properties of substances, mechanical and electromagnetic oscillations, etc.). Topics related to the protection of the environment and health have been added. Computers and video recorders are used at the secondary level but they have not influenced changes in the curriculum.

Table 13.3 (continued)
Summary of Science Reform Interests in TIMSS Countries.

Country	Any initiatives in last 5 years?	Remarks
Slovenia	Yes	The science curriculum has been enriched during the past 10 years with emphasis given to environmental issues and renewable resources. A problem has been the reduction of the number of periods for science subjects in spite of strong linking with other subjects in an interdisciplinary approach. New technologies have had an impact on the teaching of science, but not directly on the curriculum. Changes are reflected on student activities and individual work. Real life problems are solved during group work and research is done in schools or in cooperation with research institutions or universities.
Spain	Yes	Some developments have had an influence on implemented curriculum, as for example, nutrition and environmental concerns. In the next few years, the General Planning Law of Educational System will allow communities to have more autonomy. Computers and videos have been put in schools but no new topics have been incorporated into the intended curriculum as a result.
Sweden	Yes	Great freedom is given to teachers. Teaching in different science subjects is supposed to support each other, even if it takes place in different lessons or by different teachers, but it doesn't necessarily mean that teaching is integrated. There is a strong emphasis on experiment and laboratory work. It is explicitly stated that calculators shall not be used. No changes have been made in the secondary science curricula during the past 10 years in chemistry, general science, or physics. A new syllabus for biology was approved by the government in 1988. Further changes are expected between 1992 and 1995. A natural science course is compulsory for all programs. Integrated science will be compulsory for the social science program. Biology, chemistry, and physics will be compulsory for the natural science program.
Switzerland	—	—
USA	Yes	Historically the rationale for science instruction was to prepare a sufficient number of scientists and engineers for the nation. Since 1983 science instructors have been urged to prepare citizens for productive lives in a society that increasingly is dependent on science and technology. As a result there is increased emphasis on earlier and more science instruction in elementary schools. There is an increasing concern with environmental issues and increased attention to the interactions among science, technology, and society. There is movement away from many topics to more in-depth instruction on a limited number of topics. One way to accomplish this is "themes," which are applicable across all science fields. Another change is moving from instruction of "facts" to hands-on doing of science. Testing is affected by this development. There is increased emphasis on using real-world settings and involving students in problem solving. A very significant reform is the development of national standards for science curricula, assessments, and teaching by the National Academy of Sciences. Technology exists increasingly in schools, but it tends to be supplemental or peripheral rather than integrated. For schools that do have technology, students are able to use telecommunications to exchange scientific data and to access scientific databases.

Figure 13.3

Textbook Space Devoted to Major Performance Expectations.

These data are similar to those in figure 3.2, but for the broadest performance expectation categories in textbooks for the upper grade of Population 2. They also reveal considerable variation.

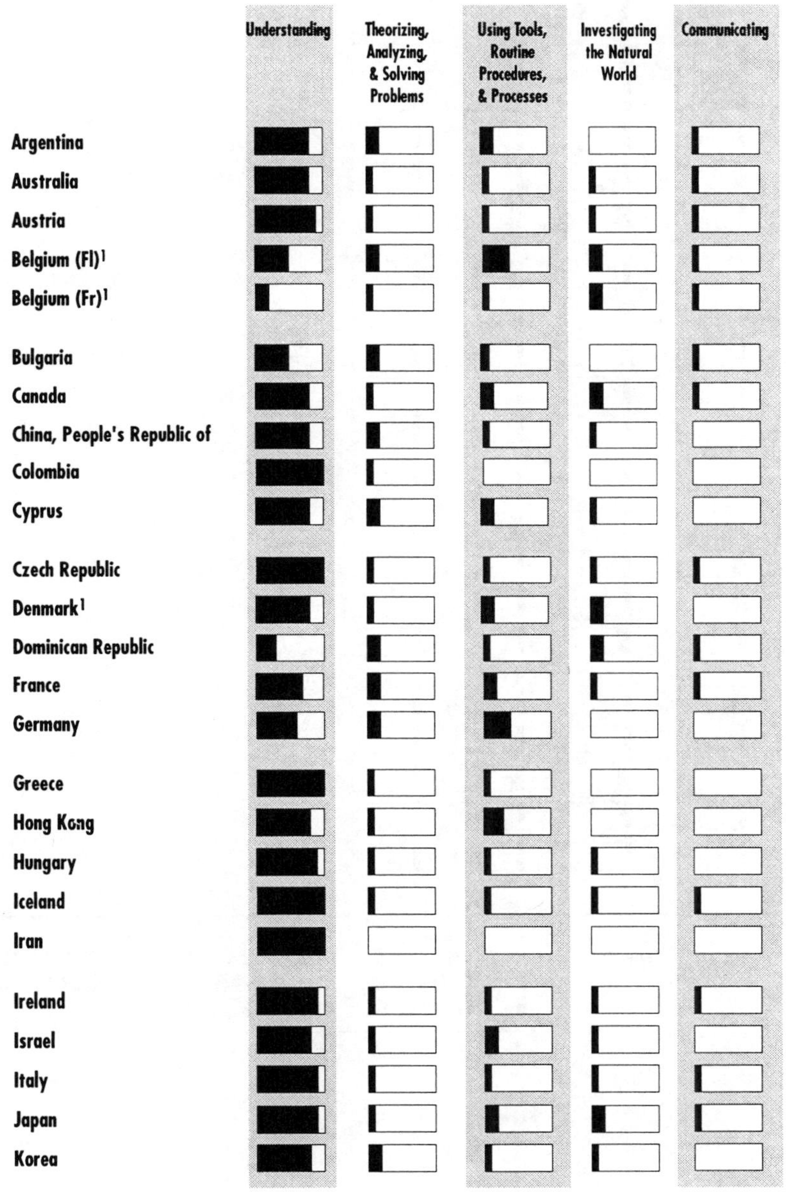

1 Refer to footnotes in figure 13.2

Figure I3.3 (continued)
Textbook Space Devoted to Major Performance Expectations.

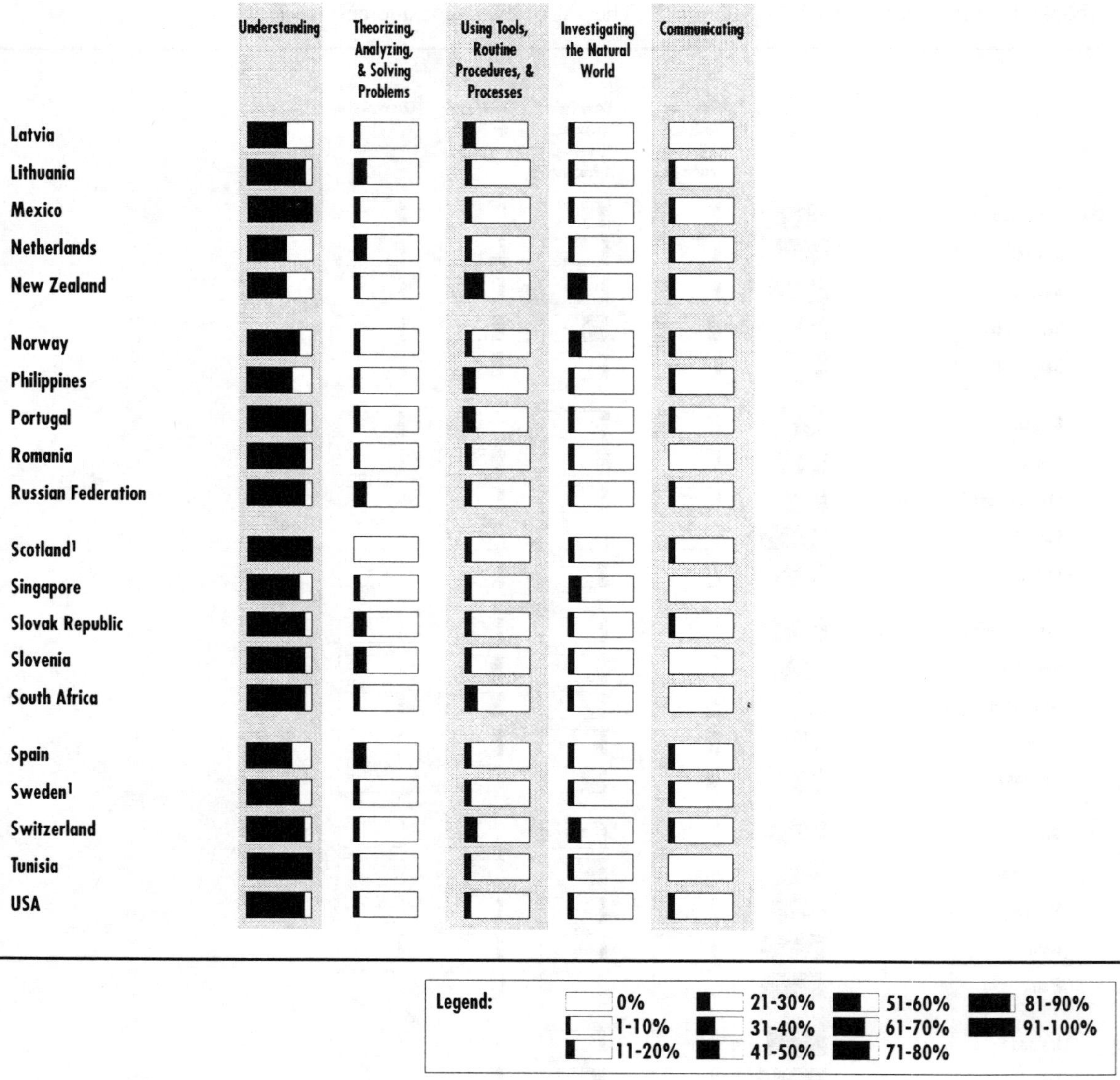

	Understanding	Theorizing, Analyzing, & Solving Problems	Using Tools, Routine Procedures, & Processes	Investigating the Natural World	Communicating
Latvia					
Lithuania					
Mexico					
Netherlands					
New Zealand					
Norway					
Philippines					
Portugal					
Romania					
Russian Federation					
Scotland[1]					
Singapore					
Slovak Republic					
Slovenia					
South Africa					
Spain					
Sweden[1]					
Switzerland					
Tunisia					
USA					

Legend:

0%	21-30%	51-60%	81-90%
1-10%	31-40%	61-70%	91-100%
11-20%	41-50%	71-80%	

Table 14.1

Specificity of Curriculum Guides.

These are the percentages for the guides of the upper grades of Populations 1 and 2 showing whether content codes and performance expectation codes were assigned at specific or more general levels of the TIMSS framework. Content codes, and, to a somewhat lesser degree, performance expectations codes were generally given at specific levels.

Country	Population 1 - Upper Grade					Population 2 - Upper Grade				
	Content			Performance Expectations		Content			Performance Expectations	
	Most Specific	Mid-Level	Most General	Most Specific	Most General	Most Specific	Mid-Level	Most General	Most Specific	Most General
Australia	82.6	15.2	2.2	94.4	5.6	81.5	18.1	0.4	84.4	15.6
Austria	60	28.6	11.4	78.1	21.9	91.6	6.6	1.9	97.8	2.2
Belgium (Fl)	74.8	2.9	22.3	84.3	15.7	83.4	10.9	5.6	98.4	1.6
Belgium (Fr)	77.5	12	10.5	97.1	2.9	83.4	11.7	4.9	85.2	14.8
Canada	89.5	4.3	6.2	98.4	1.6	94.5	5	0.5	97	3
Colombia	97.3	2.7	0	99.4	0.6	93.1	6.9	0	100	0
Czech Republic	80.4	19.6	0	100	0	97.3	2.7	0	100	0
Dominican Republic	75.7	24.3	0	-	-	25.6	73.7	0.7	-	-
France	57.9	39.5	2.6	41.7	58.3	69.7	27.1	3.2	70.5	29.5
Greece	34	59	7	86	14	68.1	31.9	0	100	0
Hong Kong	87.5	12.5	0	100	0	98.9	1.1	0	99.7	0.3
Hungary	96.8	3.2	0	100	0	96.8	3.2	0	97.4	2.6
Iceland	96	2.3	1.7	100	0	96	2.3	1.7	100	0
Ireland	95.7	4.3	0	61	39	96.1	2.7	1.1	97.4	2.6
Israel	70.5	27.1	2.4	73.3	26.7	97.9	2.1	0	97.5	2.5
Japan	85.7	14.3	0	100	0	81.1	18.9	0	68.3	31.7
Korea	100	0	0	100	0	52.9	47.1	0	65.7	34.3
Latvia	71.4	28.6	0	100	0	82.6	15.6	1.8	92.3	7.7
Mexico	86.6	13.4	0	100	0	97.2	2.8	0	99.8	0.2
Netherlands	72.3	27.7	0	93.9	6.1	75.9	22.9	1.2	89.5	10.5

New Zealand[1]	92.4	7	0.6	94.3	5.7	90.8	8.6	0.6	93.1	6.9
Norway	64.7	30.1	5.3	87.3	12.7	78.1	20.5	1.4	94.7	5.3
Portugal	89.4	10.6	0	89.4	10.6	77.4	16.2	6.4	52.6	47.4
Romania	92.9	7.1	0	100	0	72.7	27.3	0	100	0
Russian Federation	72.7	24.2	3	56.1	43.9	82	11.9	6.1	91.4	8.6
Scotland	44	26	30	74	26	44	26	30	74	26
Singapore[2]	52.9	17.6	29.4	55.7	44.3	43.8	38.4	17.8	74.6	25.4
Slovak Republic	77.3	22.7	0	100	0	84.2	9.2	6.7	80	20
Slovenia	87.4	7	5.5	98.5	1.5	87.4	7	5.5	98.5	1.5
South Africa	100	0	0	100	0	74.5	25.5	0	88.7	11.3
Spain	99	1	0	93.9	6.1	89.7	9.3	1	79.2	20.8
Sweden	62	35	3	30	70	62	35	3	30	70
Switzerland	81	17	3	99	1	78	21	1	73	27
USA	85	13	2	92	8	84	14	2	97	3

[1] Refer to footnote for figure 4.1.

[2] The national research coordinator for Singapore notes that, while considering all unit types there were a number of occasions in which the coders used mid-level or general codes, in all units devoted to the specification of content, coders used only the most specific codes.

Missing Data: Argentina, Bulgaria, People's Republic of China, Cyprus, Denmark, Germany, Iran, Italy, Lithuania, Philippines, Thailand, and Tunisia.

Figure 15.5

Specific Country/Topic Durations Compared to What Was Typical.

Statistical methods were used to produce numbers indicating how many fewer (gray-shaded icons) or more (black-shaded icons) grades the duration for a specific country and topic was from the duration for a typical country and topic — that is, a double difference. Considerable diversity is shown.

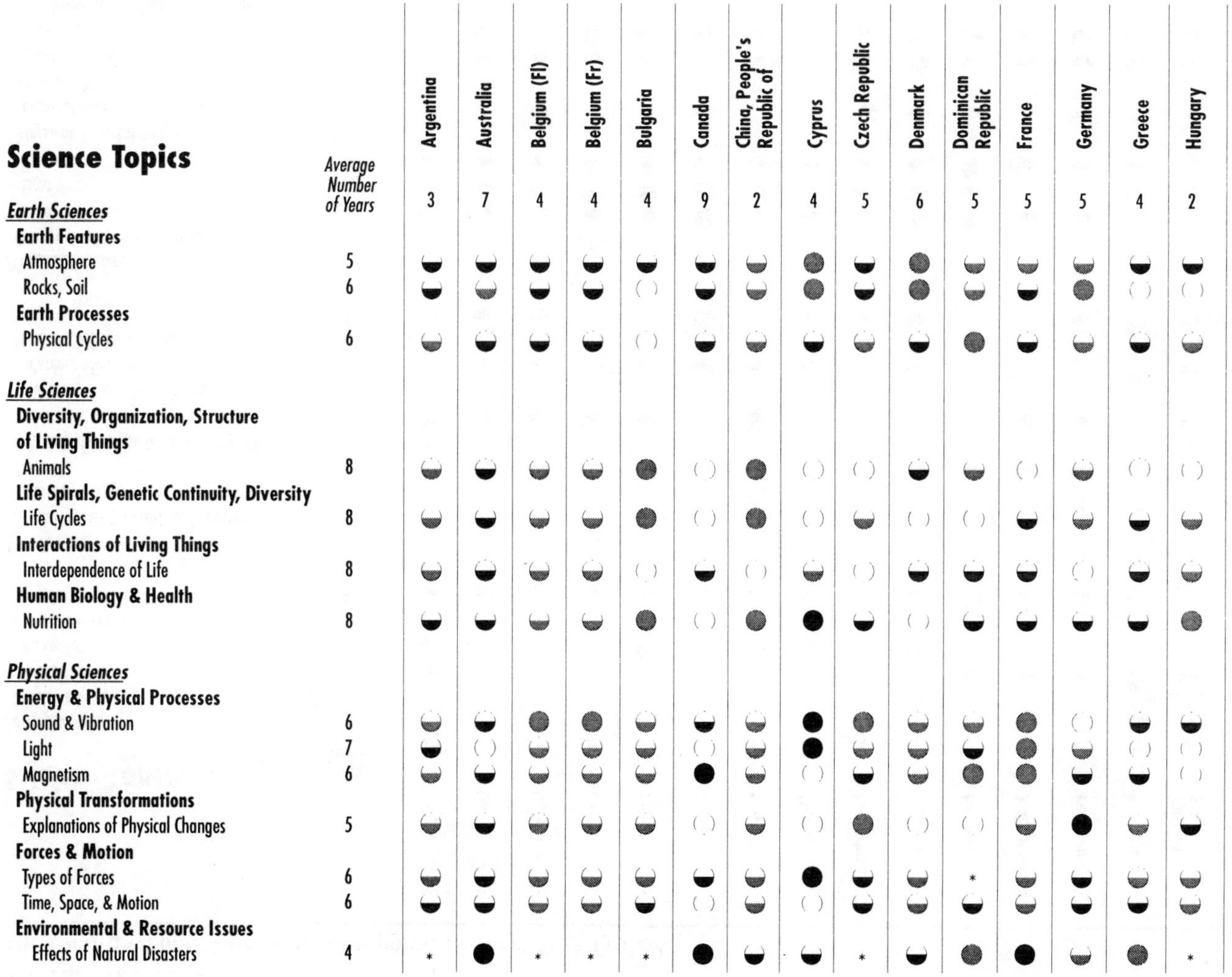

Countries (columns): Argentina (3), Australia (7), Belgium (Fl) (4), Belgium (Fr) (4), Bulgaria (4), Canada (9), China, People's Republic of (2), Cyprus (4), Czech Republic (5), Denmark (6), Dominican Republic (5), France (5), Germany (5), Greece (4), Hungary (2).

Science Topics	Average Number of Years
Earth Sciences	
Earth Features	
Atmosphere	5
Rocks, Soil	6
Earth Processes	
Physical Cycles	6
Life Sciences	
Diversity, Organization, Structure of Living Things	
Animals	8
Life Spirals, Genetic Continuity, Diversity	8
Life Cycles	
Interactions of Living Things	
Interdependence of Life	8
Human Biology & Health	
Nutrition	8
Physical Sciences	
Energy & Physical Processes	
Sound & Vibration	6
Light	7
Magnetism	6
Physical Transformations	
Explanations of Physical Changes	5
Forces & Motion	
Types of Forces	6
Time, Space, & Motion	6
Environmental & Resource Issues	
Effects of Natural Disasters	4

Figure 15.5 (continued)
Specific Country/Topic Durations Compared to What Was Typical.

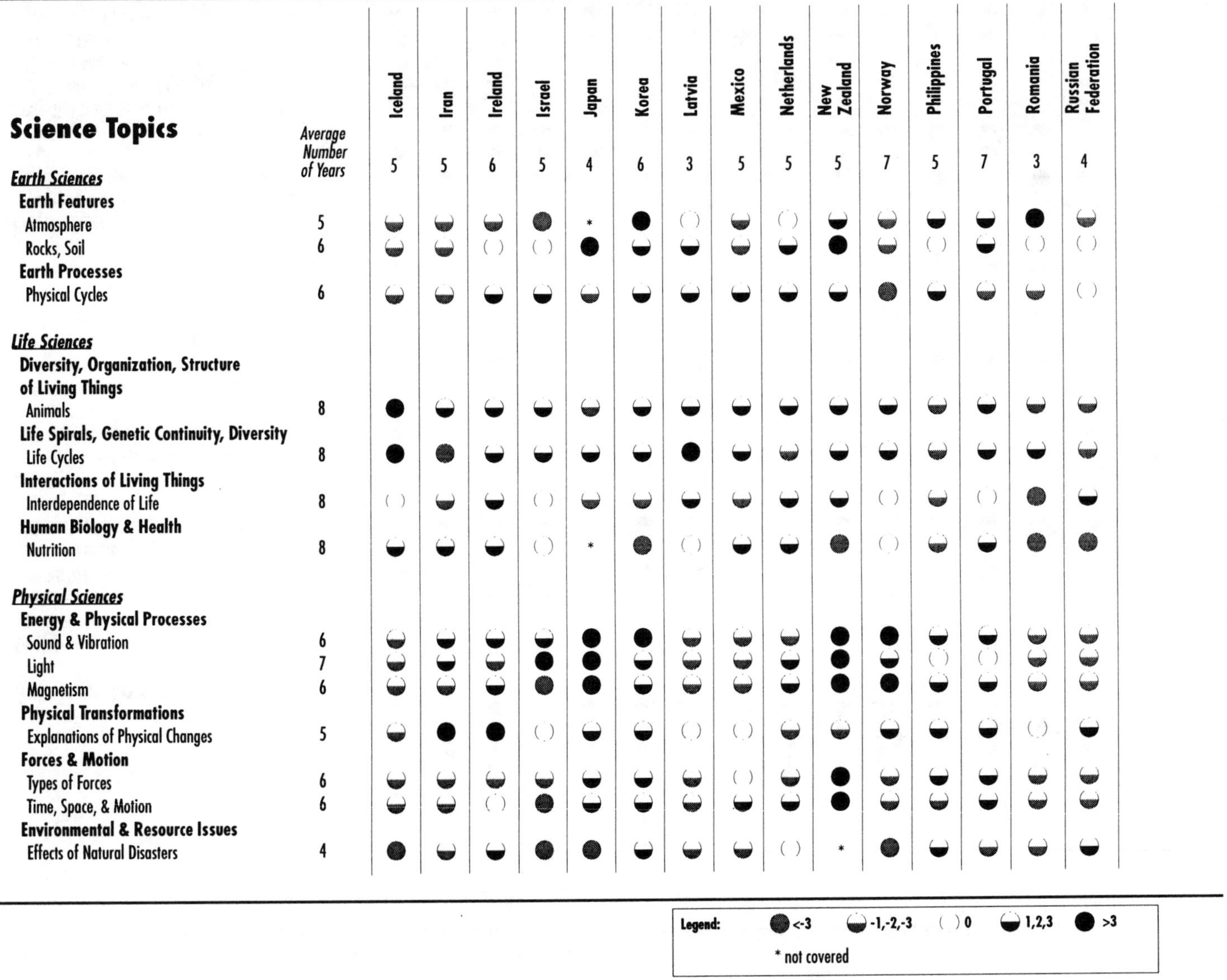

Countries (columns): Iceland, Iran, Ireland, Israel, Japan, Korea, Latvia, Mexico, Netherlands, New Zealand, Norway, Philippines, Portugal, Romania, Russian Federation

Science Topics	Average Number of Years
Earth Sciences	
Earth Features	
Atmosphere	5
Rocks, Soil	6
Earth Processes	
Physical Cycles	6
Life Sciences	
Diversity, Organization, Structure of Living Things	
Animals	8
Life Spirals, Genetic Continuity, Diversity	
Life Cycles	8
Interactions of Living Things	
Interdependence of Life	8
Human Biology & Health	
Nutrition	8
Physical Sciences	
Energy & Physical Processes	
Sound & Vibration	6
Light	7
Magnetism	6
Physical Transformations	
Explanations of Physical Changes	5
Forces & Motion	
Types of Forces	6
Time, Space, & Motion	6
Environmental & Resource Issues	
Effects of Natural Disasters	4

Country average number of years (per column): Iceland 5, Iran 5, Ireland 6, Israel 5, Japan 4, Korea 6, Latvia 3, Mexico 5, Netherlands 5, New Zealand 5, Norway 7, Philippines 5, Portugal 7, Romania 3, Russian Federation 4

Legend:
● <3   ◐ -1,-2,-3   ◑ 1,2,3   ● >3
◐ -1,-2,-3   ○ 0   ◑ 1,2,3
* not covered

Figure I5.5 (continued)

Specific Country/Topic Durations Compared to What Was Typical.

Science Topics	Average Number of Years	Singapore	Slovak Republic	Slovenia	Spain	Sweden	Switzerland	Tunisia	USA
		4	6	8	6	4	8	2	5
Earth Sciences									
Earth Features									
Atmosphere	5	0	1,2,3	1,2,3	0	-1,-2,-3	-1,-2,-3	*	1,2,3
Rocks, Soil	6	-1,-2,-3	0	0	1,2,3	-1,-2,-3	0	0	1,2,3
Earth Processes									
Physical Cycles	6	-1,-2,-3	-1,-2,-3	-1,-2,-3	1,2,3	0	-1,-2,-3	-1,-2,-3	>3
Life Sciences									
Diversity, Organization, Structure of Living Things									
Animals	8	1,2,3	1,2,3	-1,-2,-3	1,2,3	0	0	-1,-2,-3	0
Life Spirals, Genetic Continuity, Diversity									
Life Cycles	8	0	1,2,3	1,2,3	0	-1,-2,-3	-1,-2,-3	<-3	0
Interactions of Living Things									
Interdependence of Life	8	0	1,2,3	1,2,3	1,2,3	-1,-2,-3	1,2,3	-1,-2,-3	1,2,3
Human Biology & Health									
Nutrition	8	-1,-2,-3	1,2,3	1,2,3	1,2,3	1,2,3	1,2,3	1,2,3	0
Physical Sciences									
Energy & Physical Processes									
Sound & Vibration	6	0	<-3	1,2,3	1,2,3	1,2,3	1,2,3	-1,-2,-3	1,2,3
Light	7	-1,-2,-3	1,2,3	1,2,3	1,2,3	0	0	1,2,3	<-3
Magnetism	6	>3	1,2,3	<-3	1,2,3	1,2,3	0	1,2,3	<-3
Physical Transformations									
Explanations of Physical Changes	5	0	-1,-2,-3	-1,-2,-3	-1,-2,-3	-1,-2,-3	1,2,3	1,2,3	1,2,3
Forces & Motion									
Types of Forces	6	1,2,3	>3	-1,-2,-3	-1,-2,-3	-1,-2,-3	0	1,2,3	>3
Time, Space, & Motion	6	1,2,3	0	-1,-2,-3	-1,-2,-3	-1,-2,-3	1,2,3	0	>3
Environmental & Resource Issues									
Effects of Natural Disasters	4	*	*	1,2,3	-1,-2,-3	-1,-2,-3	1,2,3	*	1,2,3

Legend: <-3   -1,-2,-3   0   1,2,3   >3

* not covered

Figure 16.1
Focus Differences from the Majority.

For most topics there was one grade by which a majority of countries had focused on the topic. Individual countries had foci that differed from the majority for at least some topics. Negative numbers (gray-shaded icons) indicate focus earlier than was typical. Some differences were large enough to indicate quite different approaches to the topics involved.

Science Topics

Topic	Grade at Which the Most Countries Focused on the Topic
Earth Sciences	
Earth Features	
Rocks, Soil	9
Earth Processes	
Weather & Climate	7
Earth in the Universe	
Earth in the Solar System	7
Life Sciences	
Diversity, Organization, Structure of Living Things	
Organs, Tissues	8
Life Spirals, Genetic Continuity, Diversity	
Variation & Inheritance	12
Evolution, Speciation, Diversity	12
Physical Sciences	
Matter	
Classification of Matter	10
Structure of Matter	
Atoms, Ions, Molecules	10
Energy & Physical Processes	
Light	12
Electricity	11
Chemical Transformations	
Chemical Changes	9
Environmental & Resource Issues	
Effects of Natural Disasters	9

Columns (countries): Argentina, Australia, Belgium (Fl), Belgium (Fr), Bulgaria, Canada, China, People's Republic of, Cyprus, Czech Republic, Denmark, Dominican Republic, France, Germany, Greece, Hungary

Legend:

●	<4
◐	−1 to −4
○	0
◑	1 to 4
●	>4

* not focused   ** not covered

Note: In those instances where there is more than one focus year, the earlier one was used in creating this table.

Figure 16.1 (continued)
Focus Differences from the Majority.

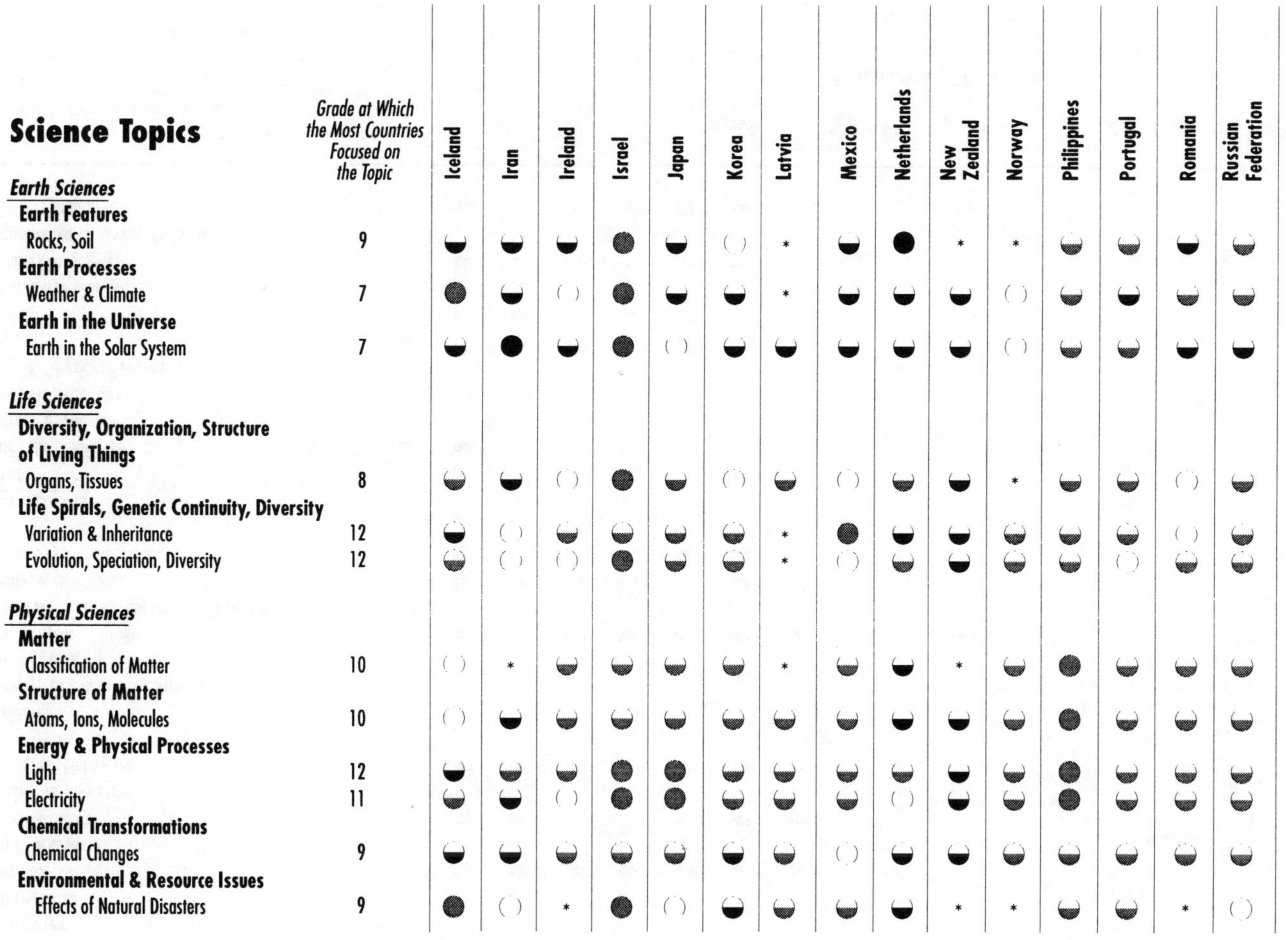

Science Topics	Grade at Which the Most Countries Focused on the Topic	Iceland	Iran	Ireland	Israel	Japan	Korea	Latvia	Mexico	Netherlands	New Zealand	Norway	Philippines	Portugal	Romania	Russian Federation
Earth Sciences																	
Earth Features																	
Rocks, Soil	9							*		●	*	*				
Earth Processes								*				○				
Weather & Climate	7															
Earth in the Universe																
Earth in the Solar System	7		●													
Life Sciences																	
Diversity, Organization, Structure of Living Things																
Organs, Tissues	8				●				○			*		○	○	
Life Spirals, Genetic Continuity, Diversity																
Variation & Inheritance	12			○				*	●							
Evolution, Speciation, Diversity	12		○					*	○						○	
Physical Sciences																	
Matter																
Classification of Matter	10							*			*		●			
Structure of Matter	10												●			
Energy & Physical Processes																
Atoms, Ions, Molecules	12				●	●							●			
Light	11															
Electricity																
Chemical Transformations																
Chemical Changes	9															
Environmental & Resource Issues																	
Effects of Natural Disasters	9	●	○	*	●	○					*	*			*	○

Figure 16.1 (continued)
Focus Differences from the Majority.

Science Topics	Grade at Which the Most Countries Focused on the Topic
Earth Sciences	
Earth Features	
Rocks, Soil	9
Earth Processes	
Weather & Climate	7
Earth in the Universe	
Earth in the Solar System	7
Life Sciences	
Diversity, Organization, Structure of Living Things	
Organs, Tissues	8
Life Spirals, Genetic Continuity, Diversity	
Variation & Inheritance	12
Evolution, Speciation, Diversity	12
Physical Sciences	
Matter	
Classification of Matter	10
Structure of Matter	
Atoms, Ions, Molecules	10
Energy & Physical Processes	
Light	12
Electricity	11
Chemical Transformations	
Chemical Changes	9
Environmental & Resource Issues	
Effects of Natural Disasters	9

Country columns (as shown, right to left): USA, Tunisia, Switzerland, Sweden, Spain, Slovenia, Slovak Republic, Singapore

Legend:
- ● <-4
- ◑ -1 to -4
- ○ 0
- ◐ 1 to 4
- ● >4
- * not focused ** not covered

Note: In those instances where there is more than one focus year, the earlier one was used in creating this table.

Figure 16.4
Coverage of Commonly Intended Topics Over Time.

The cross section of commonly intended topics at the upper grade of Population 2 is here supplemented by the curricular flow data to show whether a topic was widely covered in guides and/or textbooks, was intended to be covered before, or was intended to be covered later. This figure presents data for all countries for a sample of topics.

Science Topics — Australia, Belgium (Fl)[1], Belgium (Fr)[1], Canada, Czech Republic, Denmark[2], Dominican Republic

- Bodies of Water
- Animals
- Organs, Tissues
- Evolution, Speciation, Diversity
- Disease
- Physical Properties of Matter
- Chemical Properties of Matter
- Atoms, Ions, Molecules
- Energy Types, Sources, Conversions
- Light
- Electricity
- Explanations of Chemical Changes
- Organic & Biochemical Changes
- Applications of Science in Mathematics, Technology

Science Topics — France, Germany, Greece, Hungary, Iceland, Ireland, Israel

- Bodies of Water
- Animals
- Organs, Tissues
- Evolution, Speciation, Diversity
- Disease
- Physical Properties of Matter
- Chemical Properties of Matter
- Atoms, Ions, Molecules
- Energy Types, Sources, Conversions
- Light
- Electricity
- Explanations of Chemical Changes
- Organic & Biochemical Changes
- Applications of Science in Mathematics, Technology

[1] The National Research Coordinators of Belgium have collected data only from curriculum guides. Due to the great level of detail of the guides and their extensive use, data from these are compared in this display with the textbook data supplied from all other countries.

[2] Denmark only provided data collected from physical science textbooks for this population.

[3] The Korean NRC has requested that the presence of these topics in curriculum guides not be noted although these codes were assigned by coders to the guides provided for this population.

[4] The Korean NRC has requested that these topics be noted as present in curriculum guides although these codes were not assigned by coders to the guides provided for this population.

[5] New Zealand: TIMSS analyses were based on data from the 1989 revised edition of the curriculum guides which were valid until 1995. New guides were published in 1993 and schools were advised to work towards their implementation.

[6] In Sweden, curriculum objectives apply to groups of grades, not individual grades. Caution is suggested when interpreting these data. Though chemistry textbooks do exist for this level, none were coded.

Figure I6.4 (continued)
Coverage of Commonly Intended Topics Over Time.

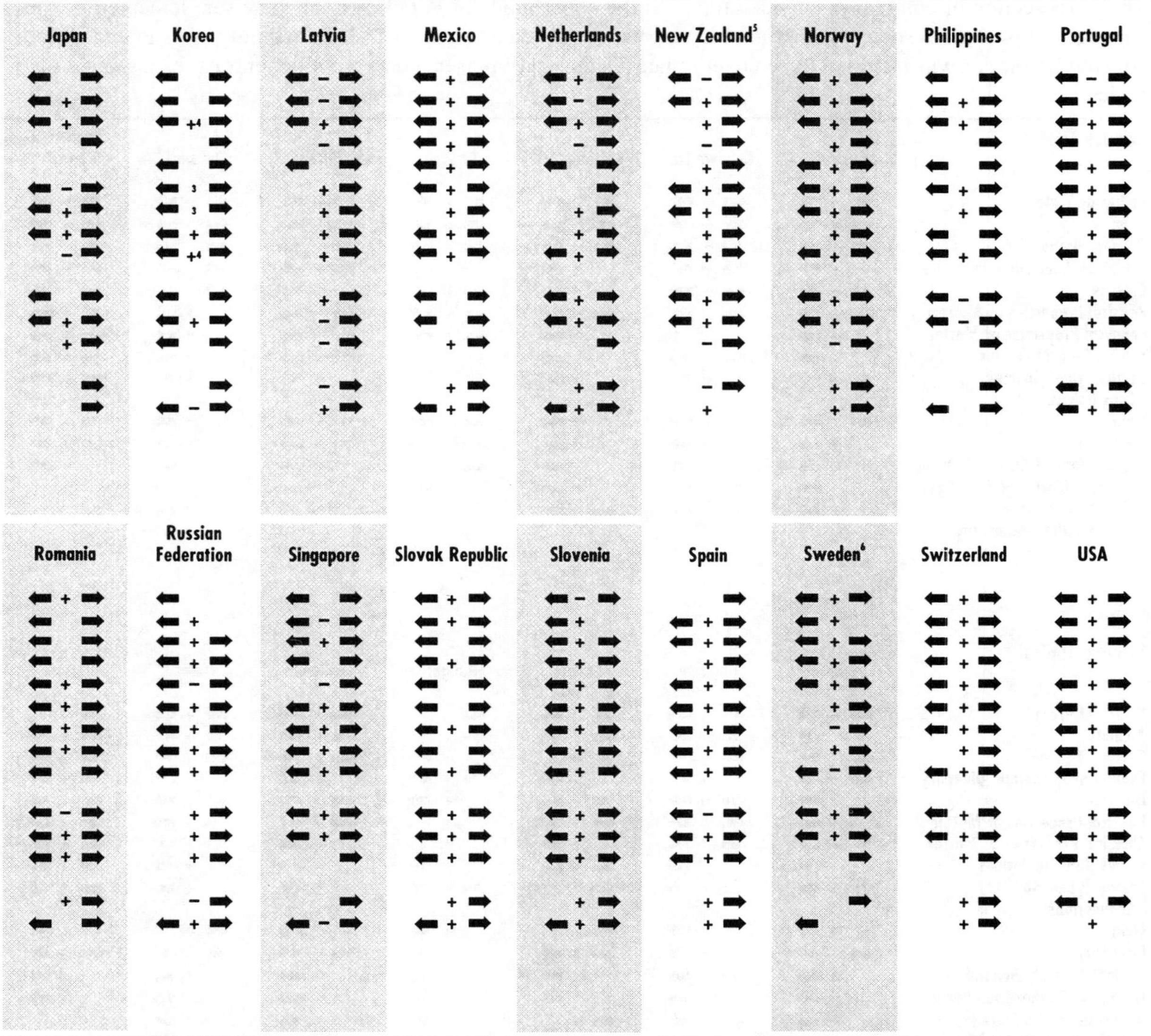

Legend:
← Intended for coverage in grades prior to the upper grade of Population 2.
→ Intended for coverage in grades after the upper grade of Population 2.
+ In both the curriculum guide and the textbook at the upper grade of Population 2.
− Present in the curriculum guide at the upper grade of Population 2.

Figure 18.1
Variation in Textbook Use of Performance Expectations - Population 1.

The textbook block percentages for various performance expectations are shown for the upper grade of Population 1. The data are for all expectations for all countries. The data again show considerable variation.

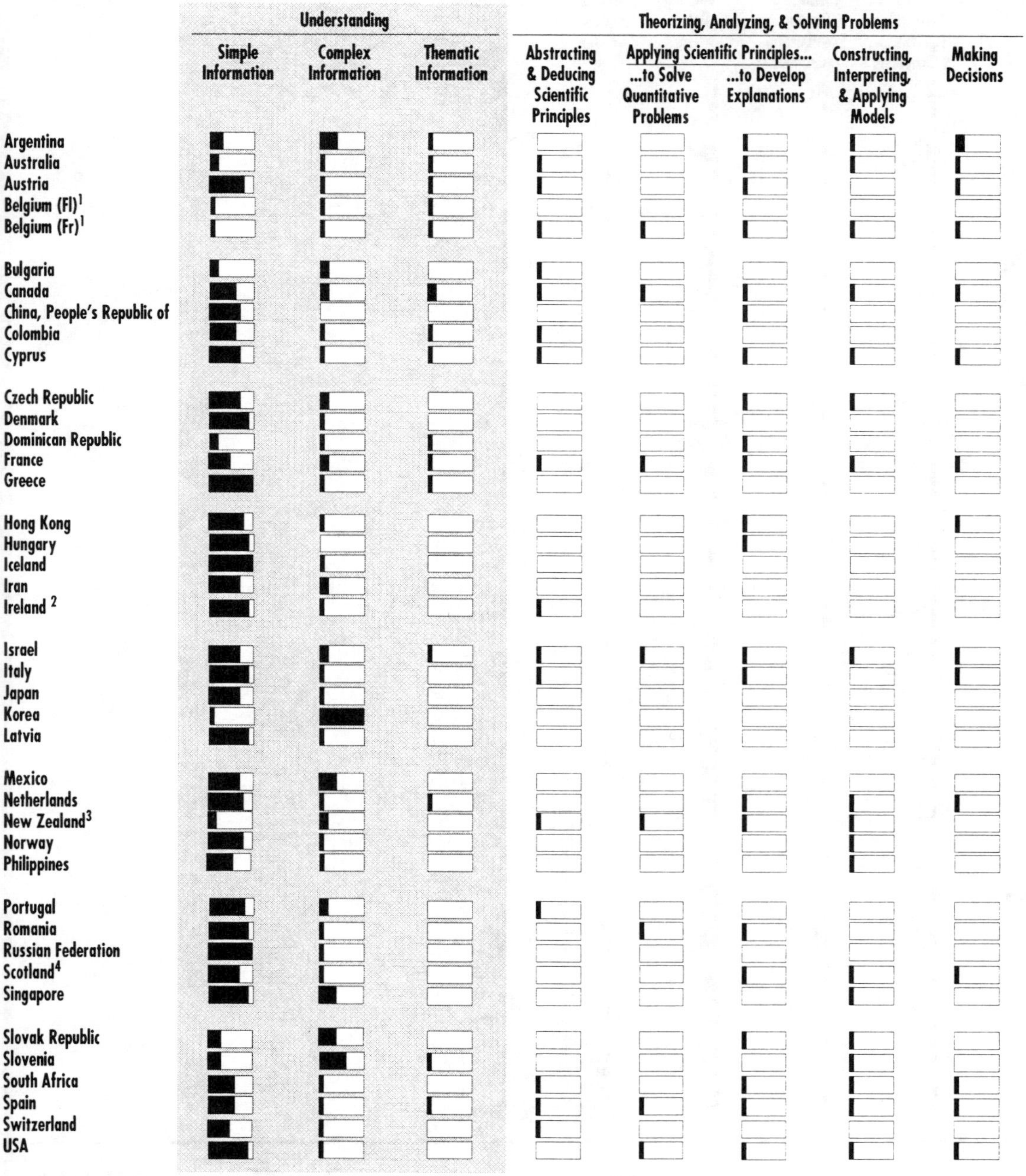

	Understanding			Theorizing, Analyzing, & Solving Problems				
	Simple Information	Complex Information	Thematic Information	Abstracting & Deducing Scientific Principles	Applying Scientific Principles... to Solve Quantitative Problems	...to Develop Explanations	Constructing, Interpreting, & Applying Models	Making Decisions

Argentina
Australia
Austria
Belgium (Fl)[1]
Belgium (Fr)[1]

Bulgaria
Canada
China, People's Republic of
Colombia
Cyprus

Czech Republic
Denmark
Dominican Republic
France
Greece

Hong Kong
Hungary
Iceland
Iran
Ireland [2]

Israel
Italy
Japan
Korea
Latvia

Mexico
Netherlands
New Zealand[3]
Norway
Philippines

Portugal
Romania
Russian Federation
Scotland[4]
Singapore

Slovak Republic
Slovenia
South Africa
Spain
Switzerland
USA

---

1   The National Research Coordinators of Belgium have collected data only from curriculum guides. Due to the great level of detail of the guides and their extensive use, data from these are compared in this display with the textbook data supplied from all other countries

2   Ireland: Data applies to lower grade of Population 1.

3   New Zealand: Since textbooks are not used in science instruction at this level, data are collected from teacher resource booklets. It is important to bear in mind that teachers use their own discretion in choosing units and topics to teach at this level.

Figure 18.1(continued)
Variation of Textbook Use of Performance Expectations - Population 1.

	Using Tools, Routine Procedures, & Science Processes				
	Using Apparatus, Equipment, & Computers	Conducting Routine Experimental Operations	Gathering Data	Organizing & Representing Data	Interpreting Data
Argentina					
Australia					
Austria					
Belgium (Fl)[1]					
Belgium (Fr)[1]					
Bulgaria					
Canada					
China, People's Rep. of					
Colombia					
Cyprus					
Czech Republic					
Denmark					
Dominican Republic					
France					
Greece					
Hong Kong					
Hungary					
Iceland					
Iran					
Ireland [2]					
Israel					
Italy					
Japan					
Korea					
Latvia					
Mexico					
Netherlands					
New Zealand[3]					
Norway					
Philippines					
Portugal					
Romania					
Russian Federation					
Scotland[4]					
Singapore					
Slovak Republic					
Slovenia					
South Africa					
Spain					
Switzerland					
USA					

Legend:
0%	21-30%	51-60%	81-90%
1-10%	31-40%	61-70%	91-100%
11-20%	41-50%	71-80%	

Figure I8.1(continued)
Variation of Textbook Use of Performance Expectations - Population 1.

	Investigating the Natural World					Communicating	
	Identifying Questions to Investigate	Designing Investigations	Conducting Investigations	Interpreting Data	Formulating Conclusions From Data	Accessing & Processing Information	Sharing Information
Argentina							
Australia							
Austria							
Belgium (Fl)[1]							
Belgium (Fr)[1]							
Bulgaria							
Canada							
China, People's Republic of							
Colombia							
Cyprus							
Czech Republic							
Denmark							
Dominican Republic							
France							
Greece							
Hong Kong							
Hungary							
Iceland							
Iran							
Ireland [2]							
Israel							
Italy							
Japan							
Korea							
Latvia							
Mexico							
Netherlands							
New Zealand[3]							
Norway							
Philippines							
Portugal							
Romania							
Russian Federation							
Scotland[4]							
Singapore							
Slovak Republic							
Slovenia							
South Africa							
Spain							
Switzerland							
USA							

Legend:
- 0%
- 1-10%
- 11-20%
- 21-30%
- 31-40%
- 41-50%
- 51-60%
- 61-70%
- 71-80%
- 81-90%
- 91-100%

Figure I8.2

Variation in Textbook Use of Performance Expectations - Population 2.

The textbook block percentages for various performance expectations are shown for the upper grade of Population 2. Similar variations as with Population 1 persisted at this level.

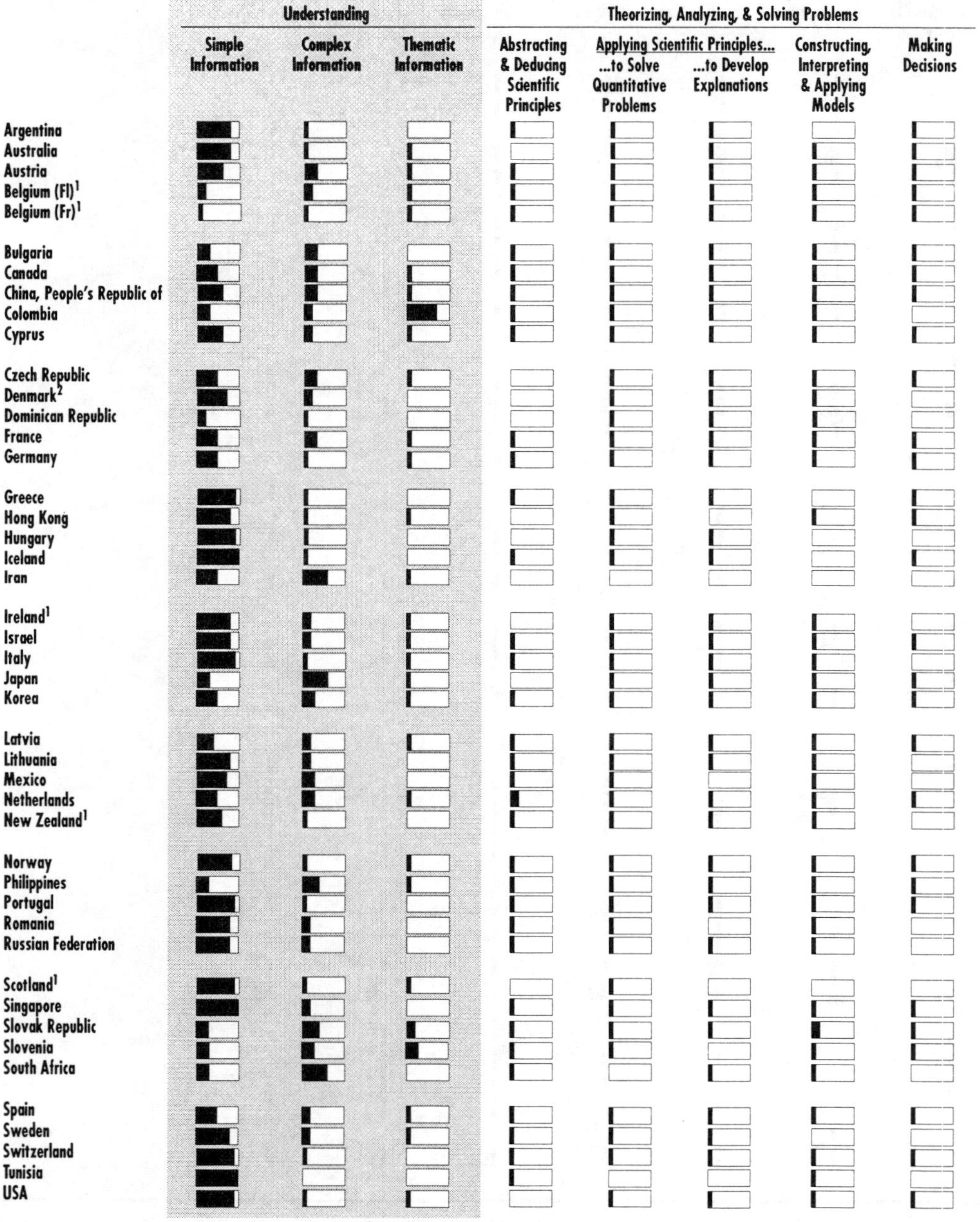

Figure I8.2 (continued)
Variation of Textbook Use of Performance Expectations - Population 2.

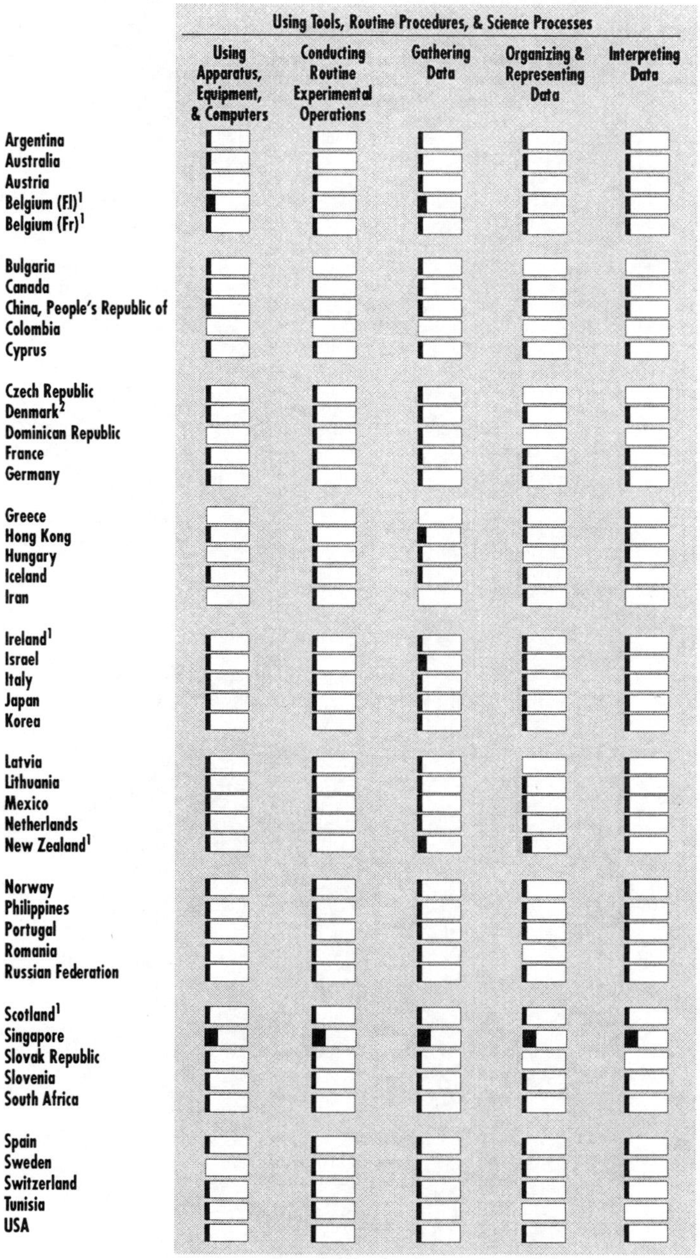

	Using Tools, Routine Procedures, & Science Processes				
	Using Apparatus, Equipment, & Computers	Conducting Routine Experimental Operations	Gathering Data	Organizing & Representing Data	Interpreting Data

1   Refer to footnotes in figure I8.1.

2   Denmark only provided data collected from physical science textbooks for this population.

Figure 18.2 (continued)
Variation of Textbook Use of Performance Expectations - Population 2.

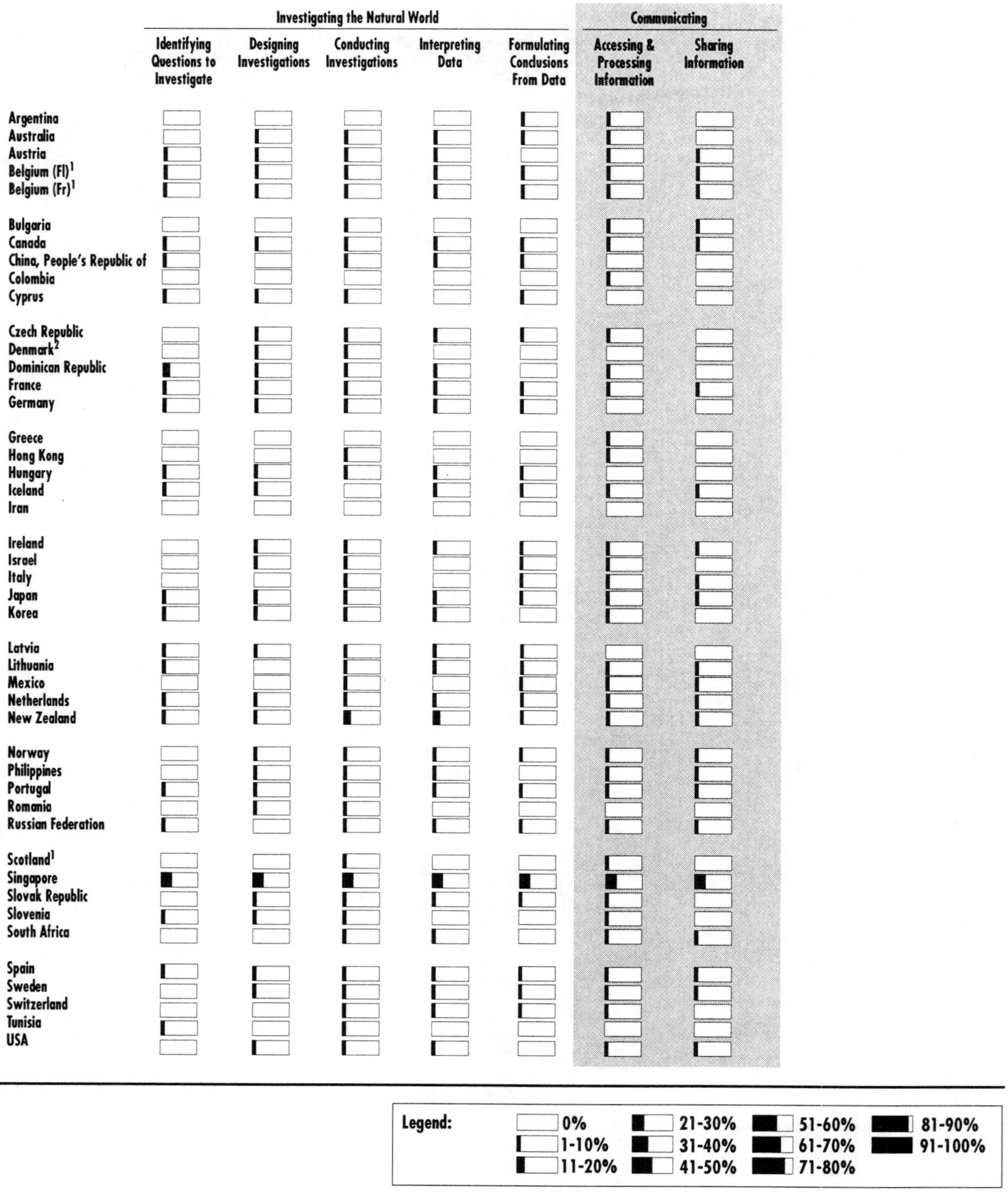

	Investigating the Natural World					Communicating	
	Identifying Questions to Investigate	Designing Investigations	Conducting Investigations	Interpreting Data	Formulating Conclusions From Data	Accessing & Processing Information	Sharing Information
Argentina							
Australia							
Austria							
Belgium (Fl)[1]							
Belgium (Fr)[1]							
Bulgaria							
Canada							
China, People's Republic of							
Colombia							
Cyprus							
Czech Republic							
Denmark[2]							
Dominican Republic							
France							
Germany							
Greece							
Hong Kong							
Hungary							
Iceland							
Iran							
Ireland							
Israel							
Italy							
Japan							
Korea							
Latvia							
Lithuania							
Mexico							
Netherlands							
New Zealand							
Norway							
Philippines							
Portugal							
Romania							
Russian Federation							
Scotland[1]							
Singapore							
Slovak Republic							
Slovenia							
South Africa							
Spain							
Sweden							
Switzerland							
Tunisia							
USA							

Legend:
0%	21-30%	51-60%	81-90%
1-10%	31-40%	61-70%	91-100%
11-20%	41-50%	71-80%	

Table 18.2

Variations in Performance Expectations.

The presence or absence of each performance expectation is indicated for the upper grades of Populations 1 and 2. These data are based only on curriculum guides. The data show that although common expectations existed across countries, they did so in a context of strong cross-national variations.

	Performance Expectation				
	Understanding			Communicating	
Country	Simple Information	Complex Information	Thematic Information	Accessing & Processing Information	Sharing Information
Australia	1/2	1/2	1/2	1/2	1/2
Austria	1/2	1/2	1/2	1/2	1/2
Belgium (Fl)	1/2	1/2	1/2	2	2
Belgium (Fr)	1/2	1/2	1/2	1/2	1/2
Canada	1/2	1/2	1/2	1/2	1/2
Colombia	1/2	-	2	1/2	1/2
Czech Republic	1/2	1/2	2	2	2
Denmark[1]	2	2	2	2	2
Dominican Republic	-	-	-	-	-
France	1/2	1/2	1/2	2	2
Germany	2	2	2	-	-
Greece	1/2	1	1	1	1
Hong Kong	1/2	-	-	1/2	1/2
Hungary	1/2	1/2	1/2	1/2	2
Iceland	1/2	1/2	1/2	1/2	1/2
Ireland	2	2	2	1/2	1/2
Israel	1/2	1/2	1/2	1/2	1
Japan	1/2	2	2	2	-
Korea	1/2	1/2	1/2	-	-
Latvia	1/2	2	1/2	2	-
Mexico	1/2	1/2	2	-	2
Netherlands	1/2	1/2	-	2	2
New Zealand[2]	1/2	1/2	1/2	1/2	1/2
Norway	1/2	1/2	-	1/2	1/2
Philippines[1]	2	2	2	-	-
Portugal	1/2	1/2	1/2	1/2	1/2
Romania	1/2	-	-	-	-
Russian Federation	1/2	1/2	1/2	1/2	1/2
Scotland	1/2	1/2	1/2	1/2	1/2
Singapore	1/2	1/2	1/2	1/2	1/2
Slovak Republic	1/2	1/2	2	-	-
Slovenia	1/2	1/2	1/2	1/2	1/2
South Africa	2	2	1/2	2	1/2
Spain	1/2	1	1	1/2	1/2
Sweden[2]	1/2	1/2	1/2	1/2	1/2
Switzerland	1/2	1/2	1/2	1/2	1/2
USA	1/2	1/2	1/2	1/2	1/2

1 Only Population 2 curriculum guides were submitted.
2 Refer to footnotes in figure 4.1.
Missing Data: Argentina, Bulgaria, People's Republic of China, Cyprus, Iran, Italy, Lithuania, Thailand, and Tunisia

Table 18.2 (continued)
Variations in Performance Expectations.

	Performance Expectation				
	Theorizing, Analyzing, & Problems				
Country	**Abstracting & Deducing Scientific Principles**	**Applying Scientific Principles... ...to Solve Quantitative Problems**	**...to Develop Explanations**	**Constructing, Interpreting, & Applying Models**	**Making Decisions**
Australia	1/2	1/2	1/2	1/2	1/2
Austria	-	2	2	1/2	-
Belgium (Fl)	2	2	2	2	2
Belgium (Fr)	1/2	1/2	1/2	1/2	1/2
Canada	1/2	1/2	1/2	1/2	1/2
Colombia	2	-	1/2	1/2	1/2
Czech Republic	1	2	1/2	-	-
Denmark[1]	2	2	2	2	2
Dominican Republic	-	-	-	-	-
France	2	1/2	2	1/2	2
Germany	-	-	-	-	-
Greece	1	1/2	1	1	1
Hong Kong	2	1/2	2	2	1/2
Hungary	1/2	2	2	2	2
Iceland	1/2	-	1/2	1/2	1/2
Ireland	2	2	2	2	2
Israel	1/2	1/2	1/2	1/2	1
Japan	2	2	2	2	2
Korea	1/2	1/2	1/2	1/2	1/2
Latvia	2	2	2	2	2
Mexico	2	2	-	2	-
Netherlands	-	2	1/2	2	2
New Zealand[2]	1/2	1/2	1/2	1/2	1/2
Norway	1/2	-	-	2	1/2
Philippines[1]	2	2	2	2	2
Portugal	1/2	1/2	1/2	1/2	1/2
Romania	-	-	-	2	-
Russian Federation	2	2	2	2	2
Scotland	1/2	1/2	1/2	1/2	1/2
Singapore	1/2	1/2	1/2	1/2	1/2
Slovak Republic	1	2	1/2	-	-
Slovenia	1/2	1/2	1/2	1/2	1/2
South Africa	1/2	2	1/2	1/2	2
Spain	-	-	-	1/2	-
Sweden[2]	-	-	-	-	-
Switzerland	1/2	2	1/2	2	1/2
USA	1/2	1/2	1/2	1/2	1/2

Legend: 1 - Present in the curriculum guide for the upper grade of Population 1
                    2 - Present in the curriculum guide for the upper grade of Population 2

Table 18.2 (continued)
Variations in Performance Expectations.

Country	Using Tools, Routine Procedures, & Science Processes				
	Using Apparatus, Equipment, & Computers	Conducting Routine Experimental Operations	Gathering Data	Organizing & Representing Data	Interpreting Data
Australia	1/2	1/2	1/2	1/2	1/2
Austria	1/2	1/2	1/2	1/2	1/2
Belgium (Fl)	1/2	1/2	1/2	1/2	1/2
Belgium (Fr)	1/2	1/2	1/2	1/2	1/2
Canada	1/2	1/2	1/2	1/2	1/2
Colombia	1/2	1/2	1/2	1	1
Czech Republic	2	2	1/2	-	-
Denmark[1]	2	2	2	2	-
Dominican Republic	-	-	-	-	-
France	1/2	1/2	1/2	1/2	1/2
Germany	-	-	-	-	-
Greece	1	1/2	1	1	1
Hong Kong	1/2	1/2	1/2	1	2
Hungary	1/2	1/2	1/2	2	1/2
Iceland	1/2	1/2	1/2	-	1/2
Ireland	1/2	1/2	1/2	1/2	1/2
Israel	1/2	1/2	1/2	1/2	1/2
Japan	2	2	2	2	2
Korea	1/2	1/2	2	2	2
Latvia	2	1/2	1/2	2	1/2
Mexico	1	1/2	1/2	1/2	1/2
Netherlands	1/2	1/2	2	2	2
New Zealand[2]	1/2	1/2	1/2	1/2	1/2
Norway	1/2	1/2	1/2	1/2	1/2
Philippines[1]	2	2	2	2	2
Portugal	1/2	1/2	1/2	1/2	1/2
Romania	2	2	2	-	-
Russian Federation	1/2	1/2	1/2	1/2	1/2
Scotland	1/2	1/2	1/2	1/2	1/2
Singapore	1/2	1/2	1/2	1/2	1/2
Slovak Republic	2	2	1/2	2	2
Slovenia	1/2	1/2	1/2	1/2	1/2
South Africa	1/2	2	1/2	1/2	2
Spain	1/2	1/2	1/2	1/2	1/2
Sweden[2]	1/2	1/2	1/2	-	-
Switzerland	1/2	1/2	1/2	1/2	2
USA	1/2	1/2	1/2	1/2	1/2

1   Only Population 2 curriculum guides were submitted.
2   Refer to footnotes in figure 4.1.
Missing Data: Argentina, Bulgaria, People's Republic of China, Cyprus, Iran, Italy, Lithuania, Thailand, and Tunisia

Table I8.2 (continued)
Variations in Performance Expectations.

	Performance Expectation				
	Investigating the Natural World				
Country	Identifying Questions to Investigate	Designing Investigations	Conducting Investigations	Interpreting Investigational Data	Formulating Conclusions From Investigational Data
Australia	1/2	1/2	1/2	1/2	1/2
Austria	1/2	1/2	1/2	1/2	1/2
Belgium (Fl)	1/2	1/2	1/2	1/2	1/2
Belgium (Fr)	1/2	1/2	1/2	1/2	1/2
Canada	1/2	1/2	1/2	1/2	1/2
Colombia	2	-	-	-	-
Czech Republic	-	-	1/2	-	-
Denmark[1]	2	2	2	2	2
Dominican Republic	-	-	-	-	-
France	1/2	1/2	1/2	1/2	1/2
Germany	-	-	-	-	-
Greece	1	1	1	1	1
Hong Kong	-	2	1/2	2	2
Hungary	1/2	2	2	1/2	2
Iceland	1/2	1/2	1/2	1/2	1/2
Ireland	1/2	1/2	1/2	1/2	1/2
Israel	1	1	1/2	1/2	1/2
Japan	1/2	2	2	2	2
Korea	-	-	1	1	-
Latvia	2	1/2	1/2	1/2	1/2
Mexico	2	-	-	1/2	-
Netherlands	1/2	1/2	1/2	1/2	1/2
New Zealand[2]	1/2	1/2	1/2	1/2	1/2
Norway	1/2	1/2	1/2	1/2	1/2
Philippines[1]	-	-	-	-	-
Portugal	1/2	1/2	1/2	1/2	1/2
Romania	-	-	2	-	-
Russian Federation	2	2	2	2	2
Scotland	1/2	1/2	1/2	1/2	1/2
Singapore	1/2	1/2	1/2	1/2	1/2
Slovak Republic	-	-	1/2	-	-
Slovenia	1/2	1/2	1/2	1/2	1/2
South Africa	2	2	1/2	1/2	1/2
Spain	-	1	1	1	-
Sweden[2]	1/2	1/2	1/2	1/2	1/2
Switzerland	1/2	1/2	1/2	1/2	1/2
USA	1/2	1/2	1/2	1/2	1/2

1   Only Population 2 curriculum guides were submitted.
2   Refer to footnotes in figure 4.1.
Missing Data: Argentina, Bulgaria, People's Republic of China, Cyprus, Iran, Italy, Lithuania, Thailand, and Tunisia

Legend:	1 - Present in the curriculum guide for the upper grade of Population 1
	2 - Present in the curriculum guide for the upper grade of Population 2

Figure 18.6

Variation in Textbook Use of Performance Expectations - Physics Specialists.

The textbook block percentages for various performance expectations are shown for books used by the physics specialists of Population 3. The data are shown for all performance expectations. A general emphasis on understanding simple and complex information was seen in many countries, as well as considerable emphasis on solving quantitative problems.

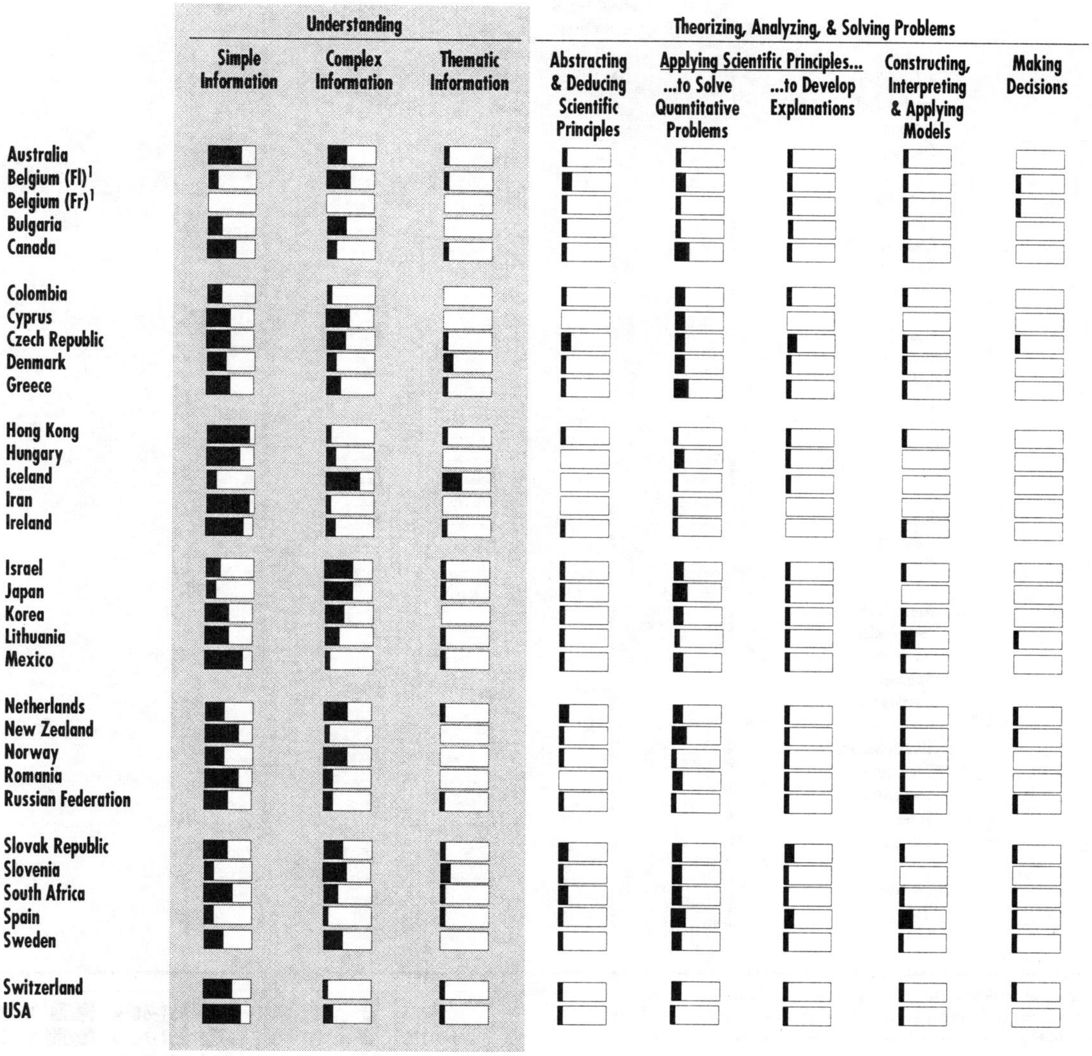

Figure I8.6 (continued)
Variation in Textbook Use of Performance Expectations - Physics Specialists.

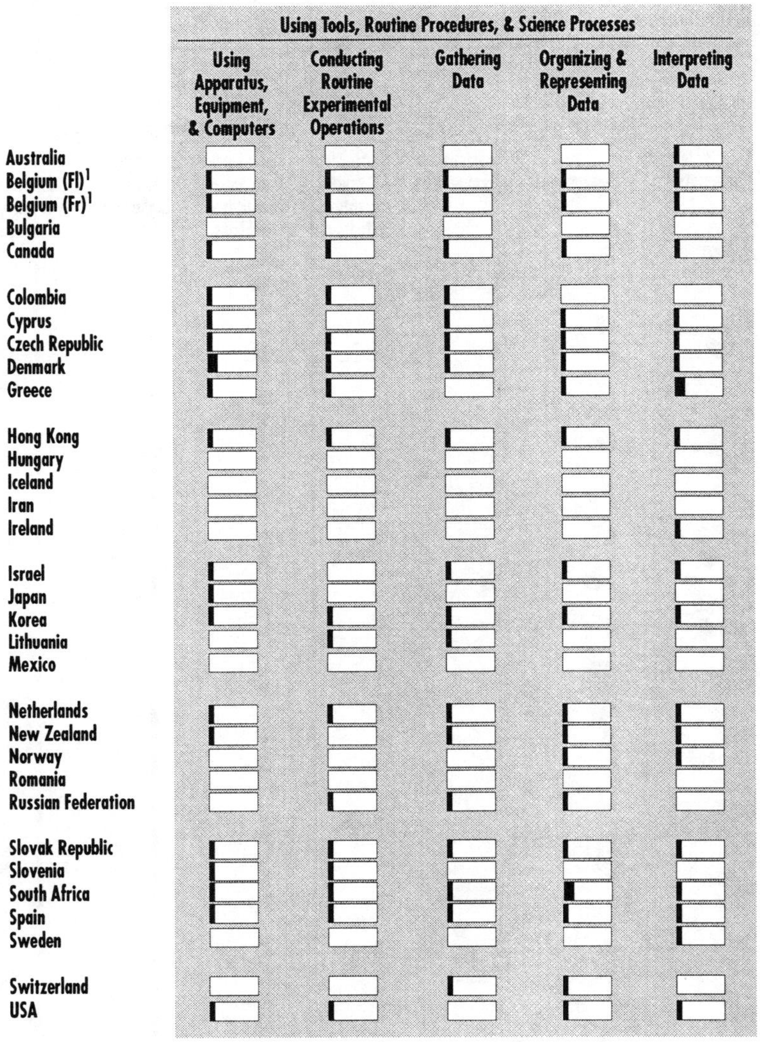

	Using Tools, Routine Procedures, & Science Processes				
	Using Apparatus, Equipment, & Computers	Conducting Routine Experimental Operations	Gathering Data	Organizing & Representing Data	Interpreting Data
Australia					
Belgium (Fl)[1]					
Belgium (Fr)[1]					
Bulgaria					
Canada					
Colombia					
Cyprus					
Czech Republic					
Denmark					
Greece					
Hong Kong					
Hungary					
Iceland					
Iran					
Ireland					
Israel					
Japan					
Korea					
Lithuania					
Mexico					
Netherlands					
New Zealand					
Norway					
Romania					
Russian Federation					
Slovak Republic					
Slovenia					
South Africa					
Spain					
Sweden					
Switzerland					
USA					

1   The National Research Coordinators of Belgium have collected data only from curriculum
    guides. Due to the great level of detail of the guides and their extensive use, data from
    these are compared in this display with the textbook data supplied from all other
    countries.
Missing Data: Argentina, Australia, People's Republic of China, Dominican Republic, France,
Germany, Italy, Latvia, Philippines, Portugal, Scotland, Slovenia, and Tunisia.

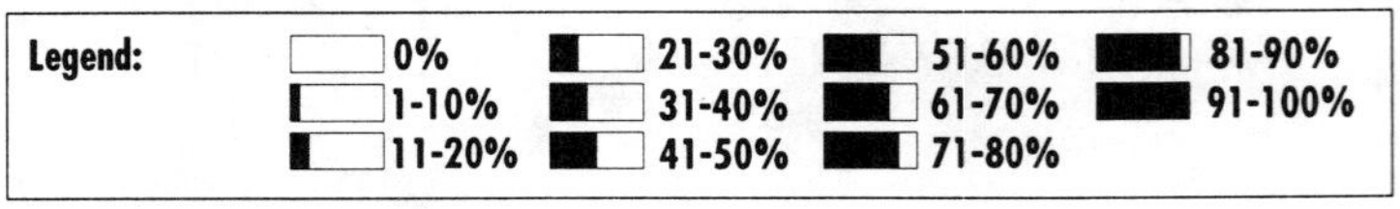

Legend:

☐ 0%	■ 21-30%	■ 51-60%	■ 81-90%
☐ 1-10%	■ 31-40%	■ 61-70%	■ 91-100%
■ 11-20%	■ 41-50%	■ 71-80%	

Figure I8.6 (continued)

Variation in Textbook Use of Performance Expectations - Physics Specialists.

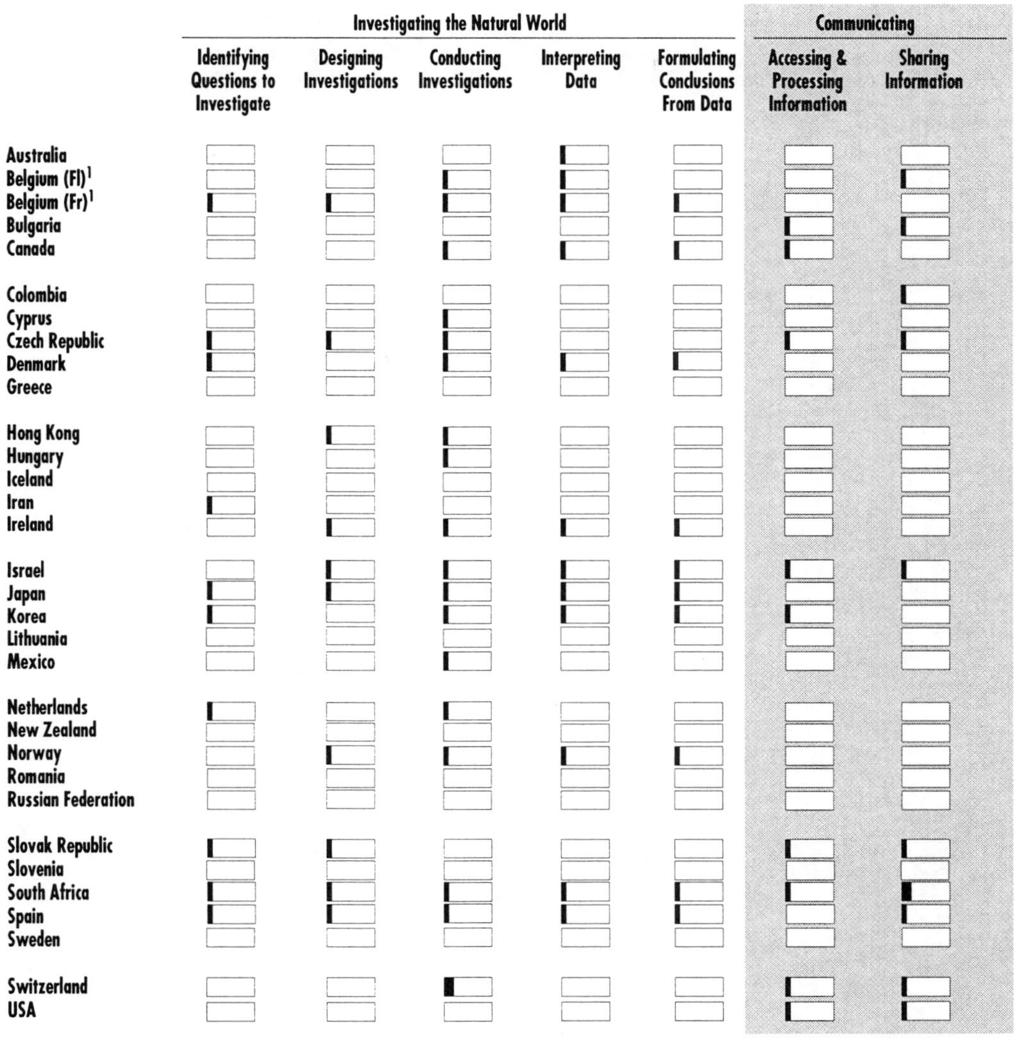

	Investigating the Natural World					Communicating	
	Identifying Questions to Investigate	Designing Investigations	Conducting Investigations	Interpreting Data	Formulating Conclusions From Data	Accessing & Processing Information	Sharing Information

Australia
Belgium (Fl)[1]
Belgium (Fr)[1]
Bulgaria
Canada

Colombia
Cyprus
Czech Republic
Denmark
Greece

Hong Kong
Hungary
Iceland
Iran
Ireland

Israel
Japan
Korea
Lithuania
Mexico

Netherlands
New Zealand
Norway
Romania
Russian Federation

Slovak Republic
Slovenia
South Africa
Spain
Sweden

Switzerland
USA

1 The National Research Coordinators of Belgium have collected data only from curriculum guides. Due to the great level of detail of the guides and their extensive use, data from these are compared in this display with the textbook data supplied from all other countries
Missing Data: Argentina, Australia, People's Republic of China, Dominican Republic, France, Germany, Italy, Latvia, Philippines, Portugal, Scotland, Slovenia, Tunisia.

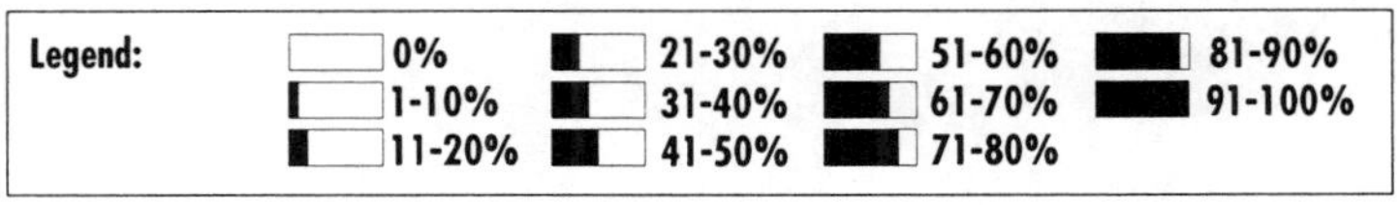

Legend:
0%
1-10%
11-20%
21-30%
31-40%
41-50%
51-60%
61-70%
71-80%
81-90%
91-100%

Appendix J

LIST OF TABLES: SCIENCE

Table 3.1: Differences in National Education Systems. This table presents information on which grades were clustered together, on the numbers of years of compulsory education and years of education offered, and on school entry ages in various countries. Cross-national differences are clear.

Table 3.2: Organizational Differences in Five Typical Countries. This display of how grades were grouped in phases of schooling in five countries clearly illustrates the strong organizational differences among countries. Indicating key grades for TIMSS achievement testing makes it clear that these differences have implications for testing.

Table 3.3: Summary of Science Reform Interests in a Representative Set of Countries. Widespread interest was expressed in science education reforms. This table summarizes reform interests in a representative sample of TIMSS countries as provided by experts within each country.

Table 4.1: Specificity of Curriculum Guides. These are the percentages for the guides of the upper grades of Populations 1 and 2 in a representative sample of countries showing whether content codes and performance expectation codes were assigned at very specific or more general levels of the TIMSS framework. Content, and to a somewhat lesser degree performance expectations, were generally given at specific levels.

Table 5.1: Introducing, Focusing On, and Completing Science Topics. There were important differences in the grade at which different science topics were introduced. There were also differences in the typical grade for completing topics and for focusing special attention on them.

Table 5.2: Science Topics Intended for Introduction in Various Stages. Certain framework topics were typically introduced at various stages of schooling: grades 1 to 3, 4 to 6, 7 to 8, 9 to 12. This table presents a broad picture of the typical sequence of science topics introduction (representing the aggregate and not individual countries).

Table 6.1: Commonly Intended Science Topics. Here are lists of the most commonly intended (at least 70 percent of countries) science topics at the upper grades of Populations 1 and 2 based on an analysis of curriculum guides and textbooks. Some topics were common only in curriculum guides, others only in textbooks, and still others in both. A few were not only included but were also emphasized in textbooks.

Table 6.2: Commonly Intended Topics for Physics Specialists. Here are the common topics for physics specialists at the end of secondary school. These topics were in the curriculum guides and/or textbooks of at least 70 percent of the countries; those indicated with an asterisk were also emphasized in textbooks.

Table 7.1: Topics Present in Both Curriculum Guides and Textbooks. Topics present in both the curriculum guides and textbooks for a given grade level in a country probably received more emphasis than those not present in both. The data for all topics and all countries are shown here for the upper grade of Population 2. Few topics received unanimous emphasis in all countries, but many were widely emphasized.

Table 8.1: Commonly Intended Performance Expectations. Performance expectations indicated in the curriculum guides of 70 percent of TIMSS countries, 70 percent of countries' textbooks, or both are shown here. Those emphasized in textbooks (accounting for 6 percent or more of textbook blocks) are marked with an asterisk. These data are only for upper grades of Populations 1 and 2.

Table 8.2: Variations in Performance Expectations. The presence or absence of each performance expectation is indicated for the upper grades of Populations 1 and 2. These data are based only on curriculum guides. The data show that although common expectations existed across countries, they did so in a context of strong cross-national variations.

Table 8.3: Commonly Intended Performance Expectations - Physics Specialists. The performance expectations indicated in the curriculum guides of 70 percent of the TIMSS countries, 70 percent of countries' textbooks, or both are shown here. Expectations emphasized in textbooks (that is, accounting for 6 percent or more of textbook blocks) are marked with an asterisk. These data are only for materials related to the Physics specialists of Population 3.

Table 8.4: Variations in Perspectives. The presence or absence of each perspective is indicated for the upper grades of Populations 1 and 2. These data are based on curriculum guides. Use of perspectives was common in many of these documents.

Table 9.1: Topic Coverage in Eight Regions - Population 1. These data are the average percentage of textbook blocks for each science topic averaged over the countries in one of eight regions. They are only for the upper grade of Population 1. Notable variations among regions occur in the areas of 'energy and physical processes,' 'diversity, organization, structure of living things,' and 'earth features.'

Table 9.2: Topic Coverage in Eight Regions — Population 2. These data are similar to those in the previous table but are only for the upper grade of Population 2. In addition to regional variations on some of the same topic areas as noted for Population 1 in Table 9.1, a variety of earth, life, and physical science topics present noteworthy regional variation.

Table 9.3: Topic Coverage for Four National Income Groups — Population 1. These data are the average percentages of textbook blocks for each science topic averaged over the countries in four national income groups (determined by World Bank criteria and data). They are for the upper grade of Population 1 only.

Table 9.4: Topic Coverage for Four National Income Groups — Population 2. These data are similar to those for the previous table, but are only for the upper grade of Population 2.

Table 9.5: Topic Coverage for Eight Statistically Determined Clusters — Population 1. These data are the average percentages of textbook blocks for each science topic area averaged for each of eight statistically determined clusters of countries. Several strong differences were seen.

Table 9.6: Topic Coverage for Eight Statistically Determined Clusters — Population 2. These data are the average percentages of textbook blocks for each science topic averaged for eight statistically determined clusters of countries. The data are only for the upper grade of Population 2.

Table E.1: Unit Types in Curriculum Guides and Textbooks. Percentages of each unit type for the documents sampled are shown for all populations and for both guides and textbooks. Curriculum guides were dominated by objective and content units. Lesson units clearly dominated textbooks.

Table E.2: Distribution of Block Types. Here are the percentages of block types for the pool of sampled curriculum guides and textbooks. Curriculum guides were mainly composed of objective, content, and pedagogical suggestion blocks. Textbooks mainly contained related graphic and narrative blocks. Activities played a larger role for younger populations: exercise sets played a larger role for older populations.

Table E.3: Proportions of Single and Multiple Content Codes. The average number of content codes assigned to a block is given for each type of document and population. The percentages of single and multiple codes among all those assigned to blocks are also given. Clearly, single and double codes predominated.

Table E.4: Proportion of Single and Multiple Performance Expectation Codes. The average number of performance expectation codes assigned to a block is given for each type of document and population. The percentages of single and multiple codes are also given.

Table I3.3: Summary of Science Reform Interests in TIMSS Countries. Widespread interest was expressed in science education reforms. Here is a summary of reform interests in TIMSS countries as reported by experts within each country.

Table I4.1: Specificity of Curriculum Guides. These are the percentages for the guides of the upper grades of Populations 1 and 2 showing whether content codes and performance expectation codes were assigned at specific or more general levels of the TIMSS framework. Content codes, and, to a somewhat lesser degree, performance expectations codes were generally given at specific levels.

Table I8.2: Variations in Performance Expectations. The presence or absence of each performance expectation is indicated for the upper grades of Populations 1 and 2. These data are based only on curriculum guides. The data show that although common expectations existed across countries, they did so in a context of strong cross-national variations.

Appendix K

LIST OF FIGURES: SCIENCE

Figure 3.1: Distribution of Block Types in Science Textbooks. The percentages of selected block types are given for science textbooks for the upper grade of Population 2. The textbooks certainly differed in what they presented and, consequently, how they might be used.

Figure 3.2: Textbook Space Devoted to Major Science Topics. This figure shows the percentages of textbook blocks devoted to the eight broadest science framework content categories for the textbooks of the upper grade of Population 2 in a representative selection of countries. This shows patterns of varying emphasis that held true for other countries and grades.

Figure 3.3: Textbook Space Devoted to Major Performance Expectations for Selected Countries. These data are similar to those in figure 3.2, but for the broadest performance expectation categories. They also reveal considerable variation.

Figure 3.4: Comparative Sequences in Four Science Topics for the People's Republic of China and Iceland. Focusing only on two countries and four topics shows that the sequence of covering aspects of topics varied as well as did overall emphasis.

Figure 4.1: Proportions of Different Units in Population 1. The figure shows the percentage of pages in science curriculum guides for the upper grade of Population 1 for each unit type. Some countries were more prescriptive-that is, had higher percentages of policy, objective, and content units. Others were relatively more facilitative - that is, had comparatively higher percentages of pedagogy units.

Figure 4.2: Proportions of Different Block Types in Population 2. This figure presents science curriculum guide block type percentages.

Figure 4.3: Pedagogical Blocks within Pedagogy Units. These are the percentages of the three types of pedagogy blocks within the pool of blocks from pedagogy units for a representative sample of countries and science guides for the upper grade of Population 2. The structure of pedagogy units were varied in many countries.

Figure 5.1: Number of Topics Introduced Very Early or Very Late. Some countries introduced many topics far earlier than was typical and some topics far later.

Figure 5.2: The Eight Topics Whose Introductory Grade Varied Most. Many topics varied considerably among countries in terms of the grade in which they were introduced. These eight topics varied the most.

Figure 5.3: Average Number of Grades Intended Across Topics by Countries. Some countries had an average duration across topics that was far less than the median for all countries. Others had an average topic duration far greater than the median.

Figure 5.4: Average Number of Grades Intended for Each Topic. More time was spent on some topics (longer average duration) than others, as these averages across all countries show. Other topics were covered for shorter times on average.

Figure 5.5: Specific Country/Topic Durations Compared to What Was Typical. Statistical methods were used to produce numbers indicating how many fewer (gray-shaded icons) or more (black-shaded icons) grades the duration for a specific country and topic was from the duration for a typical country and topic — that is, a double difference. Considerable diversity is shown.

Figure 5.6: Number of Topics To Be Covered for Each Grade in Each Country. The number of topics to be covered in various grades differs considerably across countries.

Figure 5.7: Number of Topics Introduced and Dropped. Countries differed in terms of how many topics were introduced and how many were dropped at each grade. These data show how the number of topics covered in each grade were accumulated.

Figure 6.1: Focus Differences from the Majority for Selected Countries. For most topics there was one grade by which a majority of countries had focused on the topic. Individual countries had foci that differed from the majority for at least some topics. A few countries were selected to illustrate the differences in which focus differed from the majority. Negative numbers (gray-shaded icons) indicate focus earlier than was typical. Some differences were large enough to indicate quite different approaches to the topics involved.

Figure 6.2: Proportions of Countries Covering Each Topic at Each Grade. The actual percentages of countries covering each topic at each grade were used to identify a coverage level indicated by one of the four symbols. The display makes it easier to identify topics and grades for which there were strong similarities in content coverage.

Figure 6.3: Proportions of Countries Focusing on Each Topic at Each Grade. The same system as in the previous figure is used to portray topics and grades at which curricular attention was commonly focused. Far fewer common foci are seen than topics and grades with high common coverage.

Figure 8.4: Performance Expectations for 'Animals.' Data similar to those for the previous figure are presented for the next most common topic for the upper grade of Population 1. Again, considerable emphasis was placed on 'understanding simple information'.

Figure 8.5: Performance Expectations for 'Organs, Tissues.' Data similar to those in Figure 8.3 are presented on the most common topic for the upper grade of Population 2. Again, greater emphasis is shown on 'understanding simple information,' although there is considerable more emphasis in 'understanding complex information' than had been the case with the most commonly covered topics for the upper grade of Population 1.

Figure 8.6: Variations in Textbook Use of Performance Expectations - Physics Specialists. The textbook block percentages for various performance expectations are shown for books used by the physics specialists of Population 3. The data are shown for all performance expectations but for a representative selection of countries. A general emphasis on understanding simple and complex information was seen in many countries as well as considerable emphasis on solving quantitative problems.

Figure 8.7: Variations of Textbook Use of Perspectives - Population 1. The textbook block percentages for various perspectives are shown for the upper grade of Population 1. The data for all perspectives, and for all countries, are shown. A small number of countries made considerable use of perspectives.

Figure 8.8: Variations in Textbook Use of Perspectives - Population 2. Data similar to those in the previous table are shown but for the upper grade of Population 2.

Figure A.1: Textbooks — The Potentially Implementable Curriculum. Textbooks served as intermediaries in turning intentions into implementations. They helped make possible one or more potential implementations of science curricular intentions.

Figure B.1: The TIMSS Model of Potential Educational Experiences. The means by which curricular aims and choices affect what happens in classrooms, and what students attain is part of an intricate system that includes many factors.

Figure C.1: Content Categories of the Science Framework. Each aspect of the framework contained a set of main categories. Each main category contained one or more levels of more specific sub-categories. The main content categories are shown here with some sub-categories expanded to give a better insight into the framework's structure.

Figure D.1: Two Representative Topic Trace Maps: These data are typical topic trace maps for a sample of countries selected to show representative diversity among maps produced by topic trace mapping. The results are typical of those for other topics and countries.

tions. A general emphasis on understanding simple and complex information was seen in many countries, as well as considerable emphasis on solving quantitative problems.